Strategy and Policy for Trans-European Networks

Strategy and Policy for Trans-European Networks

Debra Johnson
Senior Lecturer, Department of Marketing and Business Strategy, Hull University, UK

Colin Turner
Lecturer in International Business Strategy, Hull University, UK

palgrave
macmillan

First published 2007 by
PALGRAVE MACMILLAN
Houndmills, Basingstoke, Hampshire RG21 6XS and
175 Fifth Avenue, New York, N. Y. 10010
Companies and representatives throughout the world

PALGRAVE MACMILLAN is the global academic imprint of the Palgrave Macmillan division of St. Martin's Press, LLC and of Palgrave Macmillan Ltd. Macmillan® is a registered trademark in the United States, United Kingdom and other countries. Palgrave is a registered trademark in the European Union and other countries.

ISBN-13: 978–1–4039–4283–8 hardback
ISBN-10: 1–4039–4283–8 hardback

This book is printed on paper suitable for recycling and made from fully managed and sustained forest sources. Logging, pulping and manufacturing processes are expected to conform to the environmental regulations of the country of origin.

A catalogue record for this book is available from the British Library.

Library of Congress Cataloging-in-Publication Data
Johnson, Debra, 1957-
Strategy and policy for trans-European networks / Debra Johnson, Colin Turner.
p. cm.
Includes bibliographical references and index.
ISBN-13: 978–1–4039–4283–8 (cloth)
ISBN-10: 1–4039–4283–8 (cloth)
1. European cooperation. 2. Transportation–European Union countries–International cooperation. 3. Telecommunication–European Union countries–International cooperation. 4. Power resources–European Union countries–International cooperation. I. Turner, Colin, 1967- II. Title.

JN30.J53 2007
388.094–dc22 2006051592

10 9 8 7 6 5 4 3 2 1
16 15 14 13 12 11 10 09 08 07

Printed and bound in Great Britain by
Antony Rowe Ltd, Chippenham and Eastbourne

Contents

List of Tables

List of Graphs

List of Figures

Preface

We first began to take a serious interest in TENs over ten years ago when they featured regularly as part of the solution to Europe's general economic and competitiveness ills. At that time, the emphasis was on the development of the policy instruments needed to make TENs a reality and our research interests focused on the potential contribution of a 'joined up' infrastructure to the achievement of the broad objectives of European integration. Since then, although the refining of policy instruments and the development of new instruments continues, emphasis has shifted to the construction and development of individual TENs projects.

Our research interests have moved on as well. Whilst retaining our interests in the individual TENs sectors, we have become increasingly interested in the relationship between public policy and corporate strategy – a relationship which forms the core theme of this book. Notably, in the context of TENs, this means the interplay between the policy factors influencing the development of trans-European infrastructure (and these extend far beyond the TENs programme, including liberalisation in particular) and the related changes that have occurred in corporate behaviour. Accordingly, we have developed a framework designed to help analysis of the public policy-corporate strategy interface.

Chapter 1 provides the components underpinning the framework. It explores the generic nature of networks and of network industries, setting out key elements such as the distinction between nodes, services and links – a distinction which is used throughout the book – and discusses some of the relevant network literature. It then explores the public policy framework, initially by utilising and adapting Susan Strange's ideas about states and markets, particularly the concept of secondary power structures, in international political economy. This is augmented by a generic discussion of more micro industrial policy issues. These elements are then brought together into the new framework. This framework is built around two main axes – the control-liberal axis relates to public policy and the type of government intervention demonstrated within the network whereas the strategy dimension is based on market structure and the assumption that competitive conditions play an important role in determining corporate strategy.

Chapter 2 is less generic in nature and more concerned with evolution of thinking and policy relating to infrastructure development within Europe and to the TENs programme in particular. The chapter highlights shifting regimes and approaches to infrastructure policy within Europe and mirrors Chapter 1 by linking TENs policy to other EU policies, notably to the changing approach to industrial policy. Moreover, this chapter brings TENs policy up-to-date by locating it within the broader context of the relaunched Lisbon Agenda. Just as the original TENs initiative complemented the overarching integration initiative of the Single European Market so TENs in the first decade of the twenty-first century underpins the Lisbon Agenda which, like the SEM, is among other things, intended to boost the EU's competitiveness but which is unlikely to be as successful as the SEM.

By discussing the main characteristics and assumptions of generic TENs policy, the chapter provides important background for subsequent chapters. In particular, by establishing TENs in its broader integration context, the chapter sets the stage for the later tensions between political and commercial objectives experienced within individual TENs sectors.

Considerations about the transport, telecommunications and energy sectors are contained in Chapters 3, 4 and 5 respectively. The aim of the three sectoral chapters is to examine the development of TENs policy and related policies within the framework outlined in Chapter 1. Each sector poses its own challenges and has generated its own responses to the public policies, which have largely moved towards the liberal end of the policy axis during the last 15 years.

Transport TENs have had the highest public profile of the three TENs sectors. The importance of efficient transport systems to modern economies is widely recognised. The challenge for Europe has been to transform essentially national networks into one big network that serves a market which spans 25 countries (to become 27 in 2007). This already difficult challenge is being faced at a time when mobility generally is increasing, leading to bottlenecks and congestion that ultimately undermine what Europe is trying to achieve. In other words, not only is Europe trying to integrate disparate and disjointed transport markets but it is attempting this at a time of a growing mobility gap. Liberalisation, as in the other TENs sectors, is one of the key weapons in the EU's armoury: indeed it can be argued that the success of TENs and liberalisation polices are closely entwined.

Transport is a particularly complex element of Europe's infrastructure challenge because it is composed of several modes – rail, road, air, inland waterways, and shortsea shipping. The challenge is not only to liberalise and improve the infrastructure of these modes but also to enhance the overall efficiency of Europe's transport system by enhancing the links between modes. Much progress has been made in the liberalisation of transport since the early 1990s, particularly of transport services, although the timing and speed of change has varied between modes. The emergence of trans-European transport infrastructure has been slower but service providers are responding to the broader European market place and its needs and are strongly engaged in developing pan-European transport services.

Chapter 4 is concerned with telecommunications. Liberalisation was completed relatively earlier in telecommunications than in the other TENs sectors and technology has altered the rules of the game in this sector. However, the attainment of advanced telecommunications infrastructures within the context of the Information Society is a key theme in Europe's competitiveness strategy. Unlike the other two sectors, there has been a slight retreat from the objectives of liberalisation as policy makers acknowledge that incumbents, so often the major obstacle to liberalisation, may be best placed to achieve the main objective of establishing the backbone of the Information Society.

Chapter 5 relates to energy and completes the trio of chapters about individual TENs sectors. This sector has received less publicity than transport and telecommunications within the context of TENs but in these days of rising energy prices and concerns about securing sufficient reliable sources of energy, this sector has risen once more up the policy agenda. The legal framework for full gas and electricity liberalisation is in place but concerns about uneven and inadequate implementation throughout Europe are raising concerns about the type of utilities markets that are emerging within Europe and are distorting the market signals that are supposed to facilitate the emergence of integrated energy networks. Accordingly, the Commission has been assertive in trying to achieve implementation: it has instigated an inquiry into competition in the energy markets and has threatened to use the European Court of Justice to force open markets. It is in the energy sector, particularly if security of supply concerns intensify as seems probable, that there will be the biggest clash between states and markets as member states seek to support and/or create national champions. At the time of writing, two examples of this arising from the actions of the Spanish and French governments to prevent the takeover of domestic energy firms by companies from other member states counter to basic Community principles are proving a stern test of the Commission's resolve in this area.

Chapter 6, the final chapter, returns to an overarching approach with a view to pulling out similarities and differences in the network sectors in relation to the main themes of the book. Moreover, it concludes by setting the issue of transnational networks within a much wider geographical context than Europe which is where we believe the long-term challenge lies for the development of infrastructure – a challenge that will again reside at the interface between public policy and corporate strategy.

Ultimately we believe that efficient, connected infrastructure can only contribute to the success of the European economy. Achieving this objective is far from simple and involves a highly complex interplay between public policy objectives and aspirations and the needs and willingness and/or ability to act of the private sector. This book begins the process of examining this relationship but there is much scope to develop the core themes, both in relation to individual sectors, regions and beyond the European territory.

Debra Johnson and Colin Turner
Hull, August 2006

1

An Emerging Paradigm for Network Development

As an issue in the study of international business, the emphasis on the study of networks has tended to focus upon the use of networks as an organisational form or response to opportunities presented by the internationalisation process (for a review see De Man, 2003). International business research has yet, however, to recognise that networks as an organisational form depend upon physical infrastructure being in place to support their interactions across space. If the discipline of international business is concerned with the study of business activities that cross politically specified boundaries, then it must not only examine the strategic and contextual impacts of this process but has also to consider the legal and physical means of access and how that access is achieved. Thus the study of international networks is not only about the networks themselves but also concerns access to markets, both legally and physically, and the process by which this is achieved. In other words, relational networks are facilitated and made feasible by matching physical networks. This underlines that the network economy as a theme of research in international business has two interrelated components: the relational component and the physical component. The emphasis of this volume is on the development of the physical infrastructure, or networks, that enable these relational networks to function and support the internationalisation process in general.

In focusing upon the development of infrastructure networks to support the evolution of the international/global economy, it is necessary to explore the complex system through which such networks emerge. Not only do international infrastructures support the initial internationalisation process, they also help shape the corporate/business strategy, both of companies within the network sectors and of companies who utilise the facilities and services provided by the network industries themselves. The last two decades have seen a shifting paradigm in network development as the state has retreated as the primary provider and private provision has become increasingly common. This paradigm shift underlines that the development of international networks is a result of the shifting interface between state and corporate activity. Such an interface remains important even as the market-driven provision of network infrastructure matures. Despite increased private provision, a strong incentive remains for states, either passively or actively, to become involved to ensure networks develop in a manner that is compatible with the welfare of the broadly defined socioeconomic body. This reflects the fact that infrastructure fulfils many roles beyond those of the market, including public policy and international political economy roles in particular.

On the face of it, the three network sectors under examination (telecoms, energy and transport) exhibit diverse characteristics – and indeed their construction poses their own specific challenges. However, the important underlying generic characteristics they share plus the convergence, albeit limited, that is taking place between these sectors, facilitates the use of a common framework for analysis of these sectors. In particular, the network industries have the same generic structure composed of the following components:

1. Nodes/hubs: geographically fixed points on a network infrastructure that allow for the collection, co-ordination and distribution of tangible and intangible products around the network. Examples of nodes are power stations, telephone exchanges, rail stations, airports, etc.;
2. Links: the physical pathways/means of moving materials or providing services between hubs/nodes or between hubs/nodes and the end user. Links are differentiated between fixed links (such as gas pipelines, electricity transmission lines, road and rail lines, etc.) and flexible links (such as rolling stock, etc.);
3. Services: much of the value to business from a network derives directly from the services delivered across the network to the end user. These include all transport services, natural gas, electricity and all forms of telecommunication services. Traditionally the quality, price and reliability of the service were regarded as the key values associated with the network.

The aim of this chapter is to develop a framework for analysing the emergence of cross-border physical infrastructure by bringing together both the market-driven and public policy-related processes underpinning international network development. Initially, the form and nature of the network economy is examined as a means of exploring the role played by physical networks in the internationalisation process and the characteristics and impact of the network industries themselves. Public policy is then brought into the picture with a discussion at a macro level of how public policy has evolved in relation to infrastructure in general terms and how infrastructure is intertwined with issues of power and links states and markets. This is augmented with micro discussion of how industrial policy has evolved to deal with these issues on a more detailed basis. The firm and public policy strands are then brought together in a unifying framework that helps draw out the interactive relationship between state and market and between public policy and corporate strategy.

The emerging network economy and the internationalisation of infrastructure

The conventional perspective of the emerging international network economy is of commercial relations born of interconnected business relationships (Gadde *et al*, 2003). These links can be both intra- and inter-firm and shape a firm's position in the global economy in terms of its reach and position within its target markets. These are, in turn, embedded in different national environments as well as in the global business environment, the international trade system and assorted regional trade groupings. Such environments include assorted networks including social, technological, regional, infrastructure, institutional and market networks (Fletcher and Barrett, 2001). The network approach sees firms as embedded in networks of resources, information and other flows (Gulati *et al*, 2000). In turn, these can directly impact upon industry structure and on the nature of competition within it. If networks are viewed as strategic groups, they become a powerful factor influencing industry structure. Moreover, these networks become a source of value creation when they enable firms to access resources and capabilities to which they would otherwise have been excluded. The strategic benefits of these networks are derived from their structure, the nature of membership and the modality (that is, their strength and nature) of links within the network. The network can also be used to reduce transaction costs through lowering contracting and co-ordination costs.

The nature of the links varies, depending upon whether they are inter- or intra-firm. Intra-firm links reflect a conscious strategic decision by businesses as to their needs as they seek agreements with other firms to secure and enhance their position in line with their broad corporate strategy. Often the choice of how they interact with the network economy stems from increased interdependence based on an assessment of what firms feel they need to create a sustainable competitive position in an environment characterised by an increasing intensity of overseas competition. Thus in strategic terms, these networks can be used in both an offensive and a defensive manner. For MNEs, the links and flows between the constituent parts of the business are the key to its existence and survival. Such flows consist of products, resources, financial and human resources, values and culture (see Schmid *et al*, 2002). They occur because of the dispersed nature of MNE activities and are driven by the dependence or interdependence between the constituent units.

Whatever perspective one takes of internationalisation (whether it is the explicit use of networks; the gradual, evolutionary approach of the stages model; the revolutionary approach of the 'born global' model or the approach towards multinationals taken by Buckley and Casson (1976), etc.), what binds these approaches together is the need for an integrated, efficient, coherent network infrastructure to support the process. The nature of and the intensity of infrastructure usage will vary according to the degree of internationalisation but all models and explanations underline the important facilitative role in internationalisation performed by network infrastructure. Interactions and commercial activity across space needs a series of supporting physical infrastructures. This infrastructure and the services provided over it represent a cost that all businesses must absorb. The state of the infrastructure is also a determinant of cost as well as an impediment or boost to further activity.

Physical networks – be they transport, energy or telecommunications networks – have always been central to the success of business and, consequently, of economies (Nijkamp *et al*, 1994; European Roundtable of Industrialists, 1990). From a logistical perspective, businesses need networks to operate effectively and efficiently (Jarillo, 1993). As highlighted in Figure 1.1 – which offers a simplified value chain perspective of the role of networks – infrastructure networks are clearly pivotal in terms of getting the inputs 'in' to the enterprise and outputs 'out' as well as aiding internal resource mobility (especially important in the case where a business is located over several sites). The complexity of the interlinkages with other firms and resource flows between them grows evermore complex when analysis starts to focus in on value networks. Thus the role of infrastructure in product and firm development

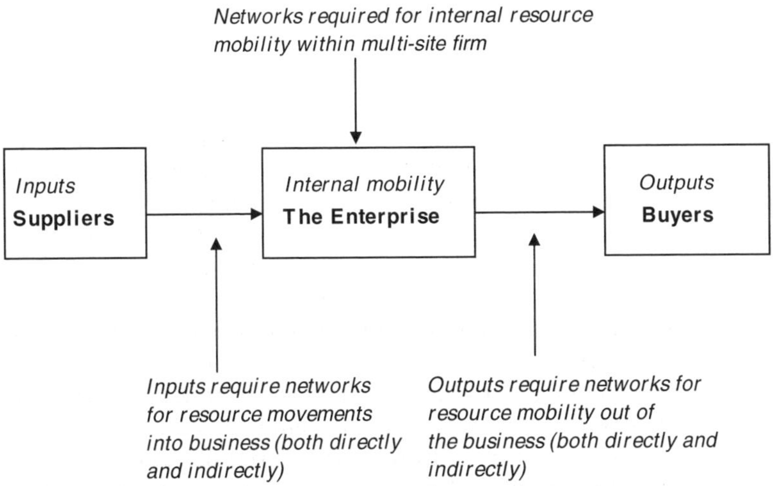

Figure 1.1 A value chain representation of the importance of networks
Source: Derived from Johnson and Turner (1997).

may not necessarily be linear but multi-faceted involving a multitude of horizontal and vertical linkages. In such a situation, the demands of physical infrastructure grow accordingly as the form and nature of the intensity of interactions grow across them.

Figure 1.1 does not need to be a multinational enterprise. However, even for a firm that is located entirely within one country, cross-border infrastructure is extremely important if inputs are imported and/or the outputs are exported. Moreover, once a firm embarks upon the internationalisation process, the existence of cross-border infrastructure links takes on even greater salience, especially given the high proportion of international trade that takes place on an intra-firm basis. Within the context of internationalisation, the following infrastructure issues are important:

- The quantity and quality of networks in the home state;
- The quantity and quality of networks between the home and host/target states;
- The quantity and quality of networks in transit states;
- The quantity and quality of networks in the target states.

Such characteristics underline the centrality of networks to the internationalisation process. The free and unfettered movement of goods and services both between and within firms across political borders depends upon the existence of high quality networks. This, together with the growing need to support global mobility, underlines the emerging strategic importance at a political level of the growing interdependence between national infrastructures. These networks are important to the development of international business not merely because of their logistical importance for the mobility of both tangible and intangible products across political borders, but because they are also important as a source of knowledge and provide a focal point for the learning process inherent within internationalisation.

These trends point to an underpinning rationale for the emergence of global networks to support the globalisation of business and the assorted international activities undertaken by firms. Internationalisation increases the degree of interdependence between separate national and regional networks to form a series of global networks that support the commercial interactions of operators across space. Each sector provides a vital service and performs an essential function in the internationalisation of business and needs to be moulded to reflect the modern commercial realities of an increasingly internationalised and integrated global economy. Internationalisation in physical networks has historically tended to be an afterthought in their development as they conventionally emerged in a highly fragmented way as they were designed to meet national needs. The increased private provision of infrastructure networks is changing this perspective. This is due to the following common features relating to the form and nature of strategies and business models for infrastructure-based businesses.

- The central value proposition and business model of infrastructure-based business are established around this core asset/resource (i.e. infrastructure) and the ability to provide services over it. This still applies even though ownership of the infrastructure and provision of services over it are increasingly separate entities;
- Corporate strategy is linked to the establishment of the necessary supporting network configuration across their target market. This configuration is based on a series of spatially dispersed hubs/nodes that are interconnected *via* a series of links. These physical elements in combination facilitate the movement and management of traffic around the network;
- The commercial success of all firms within the network industries is based upon their ability to generate traffic for their network *via* pro-

vision of network services. Thus strategies in these sectors are based upon the creation of traffic flows that utilise the available capacity upon their infrastructure;
- Network services – according to Lovelock and Yip's (1996) criteria – tend to be either 'people processing services' where the user is the focus of tangible actions (such as passenger traffic, telecoms or energy consumption) or 'possession processing services' where tangible (or non-tangible) objects are subject to actions that enhances their value to users (such as freight transport or data movements);
- The network industries and the firms within them are subject to network effects. In an increasingly internationally competitive environment, these network effects can be a key competitive advantage as – within the context of internationalising business – the more extensive the network, the more attractive it is to users, thereby increasing the likelihood of traffic generation (see Allen, 1988).

This latter point has important implications for the development of international infrastructure as it suggests that user value is linked to network size. Network effects work on both the demand and supply side, each suggesting that there is value in scale. On the demand side, the more users join or are able to access a network, the greater is the incentive for other users to join. On the supply side, the larger the network, the easier it is to lower costs and prices, both through procurement benefits and by enabling the operators to spread costs over a more extensive network. The result is a positive circle of network development. Across the network sectors, the impact of network effects varies. In telecommunications and transport, user value is directly linked to the number of other users. In energy, it is linked more to the development of effective competition and meeting security of supply concerns. This underlines that in developing infrastructures there is value in seeking to maximise the number of both actual and potential users of a network. This places an emphasis upon operators seeking to maximise the size of their network through:

- Reach: maximising the number of locations where the operator has a presence;
- Penetration: seeking to maximise access to the network by the number of users within a particular location.

The value of network effects in developing infrastructure implies that emerging interdependence between states, supranational regions and the global economy adds value to the international reach/penetration of networks.

Insofar as it extends the reach of national networks and interconnects and integrates them, internationalisation is about creating a network of networks that enables services to flow between them. Network development both in terms of intensive and extensive development requires careful choices regarding the objectives of network strategy and whether commercial or non-commercial aspects should take precedence in their development. Traditionally social aspects heavily influenced the national development of networks. In more international environments, where network services are increasingly tradable across geographic frontiers, the reach of networks will be an explicitly commercial decision with local access (or intensive service delivery) determined by national or pan-national regulators.

Public policy and infrastructure

The foregoing discussion of network characteristics and effects concentrates on the optimal network conditions and essentially implicitly incorporates market-led assumptions. This assumes that in the face of freer market forces, international infrastructures will 'naturally' emerge. However this understates the complexity of the political economy systems within which such international infrastructures will be located. Although acknowledging that the state has a role to play in physical network development, network literature in the contemporary neoliberal climate tends to confine the role of the state to finding ways of meeting public service obligations and to facilitating, or smoothing the way, for private operators and the market to construct optimal networks. In short, this literature tends to lean towards the state as a remedy for market failure, where failure is defined by the inability of the market to secure the development of a network economy that reflects the broad concerns of the socioeconomic body.

In reality, the role of states in the provision of physical infrastructure has been much wider than this and has conventionally been driven by a variety of factors, including state security, economic security, public interest and wealth creation. The priority given to these and other factors varies in time and place in line with the prevailing dominant political and economic ideology and the contemporary economic situation. In short, the relationship between state and infrastructure is constantly changing with the state dominant in some eras and locations and the markets dominant in others. Currently, although market considerations are more prominent than they have been for many years, non-market factors retain a high level of saliency and public policy and state concerns retain their importance in infrastructure considerations.

The internationalisation of economies sets such debates within a broader context as there is a need to reflect on the international political economy framework within which debates about the form and development of international infrastructure networks are located. This approach recognises inherent diversities and debates within 'the social, political and economic arrangements affecting the global systems of production, exchange and distribution, and the mix of values reflected therein' (Strange, 1994: 18). Thus developing international infrastructure is much more complex than simply establishing a physical presence and seeking to respond to simple international commercial drivers. Whilst network effects and user value linked to size stress the desire for integration, there are political forces at work that inhibit these 'natural' forces, making the system more complex (and potentially more fragmented).

Such complexity is driven by differences in the relative priority given to society's values (for example, wealth, order, justice, freedom) across economies. Such differences are determined by power relationships which have a direct influence over the interaction between state and market. Strange distinguishes between two types of power – structural power and relational power – arguing that it is increasingly structural power that is the key factor in determining how things shall be done and how frameworks in which states relate to each other, to individuals and to enterprises and other organisations are shaped. This power is derived from four distinct but interrelated structures:

- *Control over security*. Infrastructure and security have long been intimately linked. Numerous examples exist (Johnson and Turner, 1997: 1–6) of the use of transport infrastructure to protect the state against external and internal attack and to promote political power and control over either internal or conquered territory. The securing of energy supplies, particularly in more modern times of mechanised warfare, has also been a priority and telecommunications and related technology have increasingly played their part. Hard security issues continue to play an important role in relation to network sectors but 'soft' security issues are increasingly coming to the forefront. In energy, for example, concerns over supply security and the availability of supplies have long played a part in determining the structure of the industry. Environmental sustainability is another 'soft' security issue which impacts on the delivery of energy and transport services over physical networks and which involves the state, at least in determining priorities and setting targets;

- *Control over production.* Arguments regarding control over production have also shifted in recent years. Across many developed states, a trend towards privatisation has given the private sector a bigger role in production, including the network sectors. This ownership shift does not mean that the state has given up its control over production entirely. Not only does the state nurture the general environment in which business operates but it also maintains a regulatory role either through competition law or sector specific regulation (Bartle *et al*, 2002, Matheu and Jeunemaître, 2002).

  An equally important change in production, according to Strange, is the shift from a production structure that serves national markets to one that serves transnational markets. Rising interdependence underlines the role of infrastructure (as mentioned above) as the support structure – whether in terms of transportation, energy or telecommunications – on which trade and transnationally configured production networks depend. This does not fit well with the traditional role of the state in supplying physical infrastructure configured around national markets and needs. This is not to say that cross-border connections have not been made: international networks based on a series on a patchwork of national networks have emerged but the state has retained a crucial role in this process and the networks remain overwhelmingly national in nature;

- *Control over knowledge.* Whilst the knowledge structure is 'the most overlooked and under-rated structure' (Strange, 1994: 119), it is a widely acknowledged factor in the competitiveness of contemporary business. Historically, technology has played a key role in communications, reinforcing the role of the state and redefining national and international markets. This remains true for the present century where the emergence of the information economy has played an increasingly prominent role. The introduction of such technologies supports the internationalisation of production structures and facilitates marketing communications over a wider area. The control over knowledge and information conveyed by such technologies is central not only to the operation of markets but also to states and to their military and security endeavours in particular;

- *Control over credit.* Finance clearly underpins the ability of both states and players in the marketplace to exercise control over the other three power structures. Effective control over security, production and information all require access to and control over sufficient funds. Given the dual role of the three infrastructures that are the subject of this volume (that is, they fulfil roles for both states and markets), finance often has been a problematic issue for infrastructure development. The criteria for private sector investment are whether it will yield an acceptable return. The state's interest in infrastructure investment may well be one of regional development or social inclusion, for example. These create scope for tension and clashes in the development of infrastructure projects.

The development of transnational transport systems, energy supply systems and telecommunications systems are shaped by these four power structures through the interactions and the network of bargains (the relational power) within them determining the options open to players. As such, networks are 'secondary power structures'. This conclusion is supported by the fact that not only are energy, transport and telecommunications economically important in their own right, through the generation of wealth and provision of employment, but they also play a significant role in oiling the wheels of the rest of the economy. All sectors rely on energy in a wide variety of ways; transportation underpins the production and marketing networks of business and ICTs in all its guises extends the marketing reach of firms, assists the internal management of organisations and provides the means by which innovative use of knowledge and information can lead to a competitive advantage for organisations.

These shifts within international political economy that alter the links between states and markets have been evident in the history of infrastructure development as ownership has tended to swing to and fro between public and private interests in key network sectors. As there has been a renewed shift to the private sector, there has been a tendency to make a distinction between services offered over the network and the network itself. This represents a phased approach to shifting ownership and control as the private sector moves initially into services with a presence in the actual grid/network itself following (if at all). In this context, the state's role falls into three broad categories:

1. Ownership: although it remains important in some transport sectors in particular and in some countries more than others, this role has

declined the most. Moreover, partial ownership as the endgame or as a transition to full privatisation is still commonplace throughout Europe in all three main network sectors;

2. Finance: in line with its ownership role, the state traditionally bore the cost of construction of infrastructure networks. The shift to private ownership has been driven not only by ideology but also by pressure on the public purse. However, transference of ownership to the private sector, although reducing the financial contribution of the public sector to infrastructure development, has far from removed it altogether and, indeed in road and rail transport, the state continues to play a major financial role;
3. Regulation: this is a multi-faceted role with the most obvious aspect being in ensuring that, at least during the period in which competition is being introduced, former public monopolies do not abuse their incumbent/dominant position. It is the strategic – in all senses of the word – nature of infrastructure that renders this role important, at least until competition is fully established in the network sectors. This regulatory role extends to include issues such as ensuring interoperability and interconnection that are essential for the creation of seamless networks.

Given the changes that have occurred in the four main power structures, the relationship between state and market has become much more complex. The increased prominence given to market pressures has occurred for a variety of reasons. First, the rise of neo-liberalism has afforded the market a much bigger role. Secondly, (see below) there has been a change of thinking about the role of natural monopolies and the potential for competition within them due to advances in economic and competition theory. Thirdly, the shift from national to transnational markets and production requires transnational infrastructure to support it. This necessitates a new challenge to infrastructure development as states move from developing infrastructure with a sole national perspective to one where cross-border links become more salient.

Whatever the reasons for the tilt to a greater role for the private sector in infrastructure development, the state still retains a significant role both in the construction and the regulation of physical infrastructure. This role has been made more complex by the emergence of the shifting strategic dimension referred to above. Internationalisation, cross-border spillovers and emerging interdependencies increase the complexity of the development process with the concerned stakeholders existing beyond the national boundaries of the state. The emerging interdependencies imply that if networks are to emerge to support the competitive position of Europe then the shifting relationships between the state and market need to be harnessed, supported and developed in a way that is compatible with the longer term strategic objectives of the socioeconomic body.

The shifting network environment: policy and corporate changes

The end point of such shifting interactions is to develop a series of infrastructures through a diverse range of sources that bolster the competitiveness of the economy. The shift in relationships between state and market indicates that there is an increased confidence in the market to deliver tangible, competitive benefits to the economy from its involvement in the network sectors. However, there is also an implicit assumption that not all needs will be addressed through market processes. Thus networks are expected to emerge through a combination of state and private activity, with the former being a reflection of a more passive industrial policy that seeks to harness the complexities of network development for the interest of the broader socioeconomic body. This requires a broad-brush industrial policy to reflect the wide impact of network development and of the large number of stakeholders involved (Federico and Forman-Peck, 1994). The end point of such a policy is to harness the interface between states and markets and the main driver of that relationship is to improve the competitive position of the affected state(s) through improving the performance of businesses within these locations *via* improvements to infrastructure. The ubiquitous importance of these infrastructures for business creates a rationale within the policy-making community for state involvement. In the era of active industrial policy that dominated post-war business-government relations, the result was heavy state involvement in and ownership of these infrastructures. By the end of the twentieth and the beginning of the twenty-first centuries, the state-market relationship had shifted towards a more passive set of industrial policies (see Table 1.1).

Federico and Foreman-Peck (1994) stress that the development of infrastructure is about 'modifying the ecology' of the commercial environment. In broad generic terms, policy actions in this context are about facilitating the necessary actions of public and private forces to support the broader competitiveness of the economy in question.

Table 1.1 Comparison of active and passive industrial policies

Active Industrial Policy	Passive Industrial Policy
• State ownership/control • State determined levels of investment • Funded through taxation system • Priorities established by central government • Single monopoly provision of services, links and nodes • Strong strategic importance attached to public service element • Networks often planned to meet universal service requirements	• Increased private ownership of most, if not all, aspects of networks • Market determined levels of investment • Regulation to ensure broad public services elements not compromised • Demand driven networks

Within this context, two types of policy actions shape policy objectives towards the network sectors:

- Generic policies: these integrate advances in infrastructures within a broad horizontal framework. Thus actions to promote their development are heavily skewed by concerns to improve competitiveness of the economy as a whole and the enterprises nested within it;
- Sectoral initiatives: these support the ability of specific networks to enhance the competitiveness of business and possibly support individual networks in the global arena. Sector specific measures include the re-regulation and the development of common standards to promote interconnection and interoperability.

Porter (1990) highlights infrastructure as one of a number of pivotal factor conditions that determine a nation's competitive advantage. These factor conditions are derived, in part, from sustained investment in the supporting infrastructure. Thus, the notion of sustenance is important for it is not simply the existence of infrastructure that determines competitiveness but the rate at which it is changed and upgraded. In part, this provided a rationale for the liberalisation of these sectors. As the state could no longer afford the sustained investment in infrastructure required, there was a need to find an alternative source of funds. The only alternative was private finance. However, the objectives of public and private sector investment clearly differ. Thus, in a more regulated environment, the flow of funds into infrastructure investment will result from the contingencies operators face, namely the internal and external contexts of their business. Thus, governments have a sustained role in facilitating conditions to ensure such investment is forthcoming from private sources through assorted passive and active policy measures.

Porter's assumption that the core determinants of a nation's and an industry's competitiveness can be almost solely determined by national conditions has been increasingly challenged. Rugman and Verbeke (1993) and Dunning (1991) both believe that a national focus on the determinants of competitiveness underplay both the emergence of interdependencies driven by economic integration at the regional level and the increasingly prominent role played by MNEs in the international economy. In sum, a full menu of the determinants of industry and national competitiveness must include the development of the network economy. This underlines the important platform played by infrastructure, both nationally and internationally, in an economy's competitiveness. In short, interdependencies make the global economy a much more complex phenomenon than that supposed by Porter's nationally focused analytical framework. This is especially true for small, open economies whose fortunes are more closely linked with the state of competitive conditions in major trading states (Davies and Ellis, 2000). This also applies to larger nations with prominent MNEs in multiple locations.

As internationalisation has become more apparent and evident interdependencies have emerged, so policy has come to recognise that the national or regional networks are part of a network of networks that need interoperability if they are to support the needs of an increasingly powerful internationalised/globalised customer base. This is important, not only for ensuring the state is able to support the logistics and operation of international business and therefore act as a magnet for FDI, but also for providing the platform for offensive international strategies. In short, the quality of infrastructure is an important passive element in shaping corporate strategies and thus the international competitiveness of the economy as a whole. Thus such policies can breed the development of business more able to compete on the global stage. As Kogut (1983) argues, sustained competitive advantage is derived from country-based capabilities, such as rate of technology innovation and organisational assets. Integral to such processes are the emergence of a set of corporate strategies that facilitate the emergence of transnational infrastructures.

The emergence of these strategies has in part been driven by a fundamental reassessment of the potential role of the private sector in infrastructure development. Two major theoretical developments stimulated this fundamental reassessment of the interface between public policy and the network sectors. The first was the emergence of contestable market theory that stated that if the key barrier to entry (sunk (irretrievable) costs) could be overcome then competition was feasible. In the network sectors, these sunk costs are especially high. The policy prescription was to separate network provision from services, making competition possible in the network sectors (Baumol *et al*, 1982). Over time, this could feed through into higher levels of infrastructure investment as the demands placed upon it increased. Often infrastructure would remain under a single ownership with rival service providers paying to use its capacity. The second important theoretical innovation was linked with Demsetz competition. In terms of the network sectors, this stated that competition is possible in what would otherwise be monopoly situations as rival companies bid in a competitive auction for monopoly rights (usually for a given period) after which the rights are once again up for competitive bidding (see Utton, 1986). These developments have reshaped the approach towards these sectors and altered the nature of state involvement within many of the network sectors.

The result of these changes was an industrial policy framework motivated by the removal of market inefficiencies that hinder international competitiveness. In the network sectors, competition and regulatory policy have become more important. As economies have liberalised, the state has withdrawn and new regulatory designs have emerged. Network services used to be regarded as an administrative function – public necessities that should be available to all citizens with liberalisation and regulatory change transforming citizens into customers. The result is that the role of the state in network industries has largely shifted from that of a provider to that of a regulator. Thus, there is still a role for state involvement in the network industries, notably in terms of:

- Preventing abuse of a dominant position;
- Securing market access;
- Generating and promoting change;
- Intervention in standards.

Network industries pose special problems for the conduct of normal competition policy, including the possibility that a monopoly structure may maximise social welfare; excessive profits and market share at market equilibrium due to the 'winner takes all' nature of strategy in network industries and that the reduction of entry barriers may not affect market structure that much given the in-built advantages enjoyed by incumbents. Thus network sectors tend to be dominated by regulation: competition is unlikely to emerge from market forces alone and, as social efficiency is desirable, because of the potential divergence between social and private benefits.

A political economy framework for network development

The development of infrastructure and strategies to exploit that infrastructure are thus determined by a complex mix of public policy-driven factors (that is, Strange's four power structures at the macro level and the more detailed considerations at the micro level) and increasingly by a range of commercial considerations emerging from the logic of a liberalised network economy. In other words, the interface between policy and strategy comes from the overlap between the firm specific resources (FSR) of network industries (namely, the ownership of infrastructure) with country specific resources (CSRs). That is, those factors that – at least in part – determine – a firm's competitiveness (firm specific resources) are also important in helping determine a country's competitiveness.

Traditionally, such overlap would have been regarded as incompatible given the broader socioeconomic concerns involved in network infrastructure provision. Thus, not only were strategies tightly constrained but these concerns were often state-owned. The liberalisation of these sectors has altered the nature of the relationship between the state and the enterprise. As Dunning (1991) highlighted, although FSRs are endogenous to firms, they are not divorced from their environmental context and they are important, not only for domestic businesses, but also for the ability of states to act as hosts for FDI.

The interface between infrastructure as a country specific resource and its role as a firm specific resource is reflected in Figure 1.2. This highlights the different perspectives that governments and enterprises have on infrastructure. The former regard infrastructure as a politico-socioeconomic asset, whereas the latter view the asset in purely commercial terms. There are, however, significant interactions between governments and enterprises to ensure that networks in the broad environmental context shape the international competitiveness of a nation.

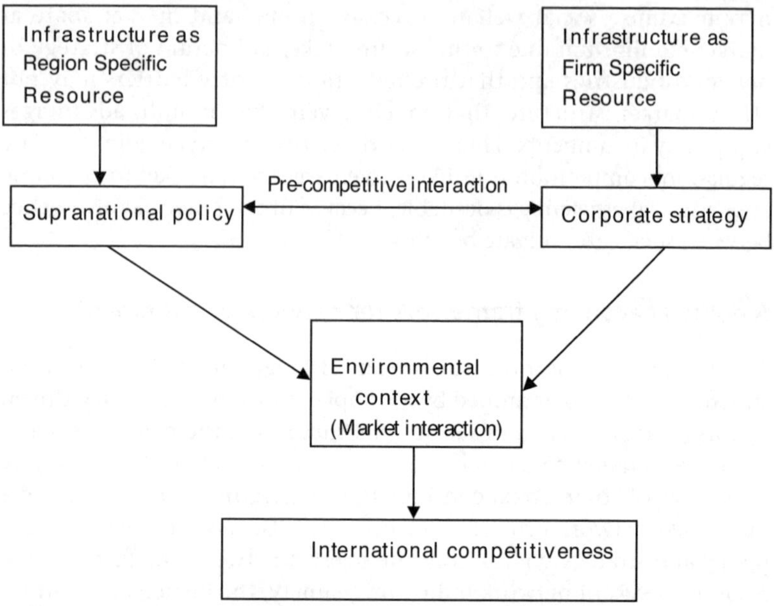

Figure 1.2 The public policy-corporate strategy interface

The framework – identified in Figure 1.2 – defines two sets of conditions in which there is interaction between government (and policy-making stakeholders) and the commercial developers of infrastructure.

- Pre-competitive interaction: this involves interaction between government bodies and network providers at the pre-competitive stage of network development. Potential interactions include direct and indirect support for research and technological development of new network technologies or support for education and training. They could also cover some of the sunk costs involved in network development;
- Market interaction: this could involve direct and/or indirect intervention in the market with or without the collaboration of the network provider. Such actions could include regulation, application of competition policy or even direct support or subsidies to ensure that broader socioeconomic objectives are met.

The nature of the interactions in the marketplace will to some extent be determined by the environmental context within which such interactions occur. Market conditions, cost issues, competitive intensity and technology, amongst other issues, will be influential over the nature and form of the interaction. If there is a more competitive market for network provision, the nature of the intervention may shift as the state may have to seek to support and interact with other providers or may seek contributions from these concerns to ensure that the broader socioeconomic concerns are met.

Figure 1.2 reflects the fact that the development of international networks have emerged as a confluence of both independent and interdependent public and private sector actions. Private sector networks have been driven by liberalisation. The national basis for transnational networks starts with the need to ensure that national networks operate as a sub-network of a broader international network. There also needs to be co-ordination between public and private actions to ensure national networks interact with other national networks and that private networks are also able to interact. These, in combination, create the international networking environment. Once this has been established, further interaction between public and private bodies at the supranational level (national governments involved) is needed to harness market process to deliver transnational networks.

As stated, infrastructure is a community resource available to all and its value increases as the number of users able to access it also increases. Mansell (1993), in examining telecommunications infrastructure, offered a bi-dimensional framework to explain network development in which two alternative infrastructure development paths are presented. The first is an ideal-type based on the competitive provision of telecommunications through a mature marketplace. The second is the strategic model which is rooted in imperfect competition and where markets are dominated by oligopolistic market structures.

The framework below develops this approach to encompass all network sectors and to underpin the interaction between policy and corporate strategy in the development of infrastructure. For the reasons outlined above, the desire for government to become involved in infrastructure has been sustained despite a trend towards increased private sector involvement in the provision of assorted parts of network infrastructure.

The different types of network strategy are differentiated according to the market structure and the form and type of intervention which is evident from the government. The type of policy-corporate strategy interface that results is a direct derivative of these two variables, see Figure 1.3.

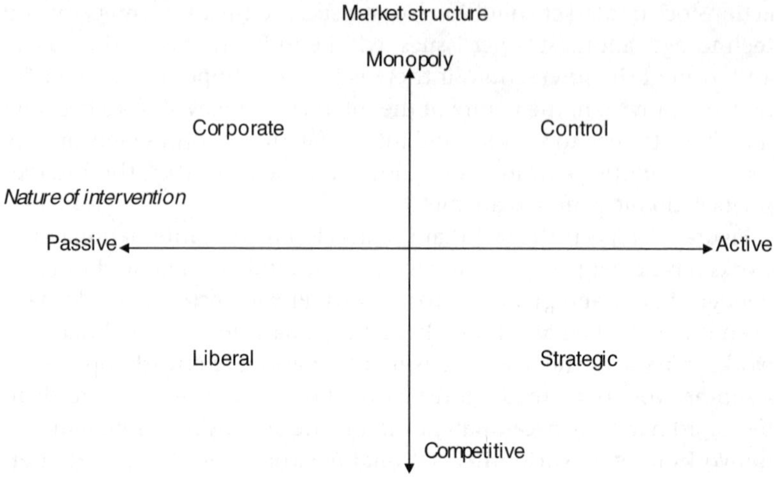

Figure 1.3 A paradigm for network development

The heterogeneity of the network sectors means that patterns of network development will vary both within and across network sectors: that is, each network sector can be subdivided into different market segments as defined by product or type of service. Levels of actual and potential competition vary according to the existence of barriers to entry. Such facets will become evident when sectors are examined in their own right.

Patterns of network development

Corporate

This form of network development occurs within the context of a passive set of policy actions where network ownership remains highly concentrated, possibly tending towards natural monopolistic conditions, and involves high fixed and often sunk costs. This type of system of network development is likely to be based on the idea that, whilst governments find some competition beneficial, excessive competition works to the benefit of no party. Too much competition may dissuade investment and defer the long-term investment needed to meet broader social priorities. Thus, government may want to limit the competition that business faces. Typically, we may be looking at Demsetz competition where a government awards a franchise to a business to enable it to run a dominant position based on regulated strategy.

The nature of these policy measures suggests that there would be little direct involvement by the state in the development of corporate strategy as a regular occurrence. Policy measures would be shaped more by ensuring that the corporate strategy of the business is not incompatible with the broader long-term social aims of the government. Thus, it is to be expected that policy imposes some constraints upon strategy *via* direct regulation in order to alleviate the possibility of market failures. These could be through direct regulation to ensure the natural monopolistic condition does not preclude consumption or simply to affect the conditions under which such products are controlled, distributed or sold. The form and nature of the interface varies according to the historical nature of the regulator and the regulated. Thus, dominance could be state-sanctioned over the short-medium term as political interests see existing dominant/concentrated market structure as the most effective method of delivering infrastructure investment. However, this does open up the possibility of regulatory capture.

A combination of political and economic inertia offers limited scope for investment and innovation beyond the incumbent operator and the state. Thus technological change takes place largely through the dominant supplier, thereby sustaining existing concentrated market structures and reflects confidence in its ability to act in a benign fashion to meet a broad set of socioeconomic concerns. Policy recognises that such control is necessary over the short/medium term due to the impossibility, or lack of will, to introduce full competition into the sector. It simply may be that competitive conditions take time to establish or that the political economy of state-sectoral relations has restrained competitive forces. Furthermore, this economic and political interface is likely to drive the internationalisation of the network. This is likely to be incumbent driven as the firm seeks access to global resource pools and international facilities. However this exposure to international markets takes place whilst there is a limited ability by the incumbent to control domestic competition (see Bonardi, 2004). This reflects the fact that the state believes that competitiveness is best addressed through regulated control until markets are capable of delivering change.

Control

Control is based on monopoly control of the system within the context of an active policy environment. This has been the traditional state of affairs for many network industries and has often involved state ownership of the provider. In this case, the interface between policy and corporate strategy is intense. Often the head of the infrastructure provider was accountable to an elected official. In this case government feels that, due to assorted reasons (such as natural monopoly or an ideological-based lack of confidence in markets), there is the need for direct control or even ownership over the means of delivery of network services. In some cases, the need for a direct interface between the operator and government may be derived from a dirigiste economic policy with the government using its ownership and control of these businesses to ensure that the strategy of the business is compatible with the broad strategic objectives of the government as a whole. This was typical in states such as France.

Under these conditions there is clearly little scope for competition and as such network innovation will tend to be state controlled. This is linked to a clear set of ideological beliefs in the benefits of state control over networks and a lack of faith in the ability of markets to deliver a solution which is deemed socially and economically desirable. This element of control is cemented by state control over finance in a sector, often through its control over the tax and revenue generating system which feeds through into infrastructure development. The nationalistic approach is also apparent in the state's influence over the form and nature of knowledge creation as well as in the design of networks for broader socioeconomic concerns. There will be a limited external dimension of these networks other than to meet access and socioeconomic concerns and is based on a system where competitiveness is best addressed through control.

The trend towards privatisation and liberalisation inevitably changes the nature of the relationship by introducing a potential conflict between the desire for control and the everyday running of the business. In this case, the government might use direct policy measures to stimulate the dominant business to act in a manner that is compatible with the broader socioeconomic aims of infrastructure. The state could use devices such as subsidies or could set a restriction on the intensity of competition as a means of meeting government objectives.

Liberal

This is based on a competitive market structure within the context of a policy framework that is generally passive. This pattern of development is pro-market and has close parallels to the idealist perspective identified by Mansell. This method seeks to achieve the objectives of infrastructure *via* the competitive actions of business. It is thus broadly demand led and based on a political faith in markets and their ability to meet broader socioeconomic concerns. In this method of development, network evolution has a close and intimate link to the strategies of operators and to their view of what is worth providing and what is not. Competitive provision extends across all aspects of the network. Network development is therefore increasingly dispersed across a range of operators.

As a direct consequence of this liberal approach, there is devolved control over security, production and finance and innovation is market sourced and driven. However, this does not preclude a degree of state finance if there is evident market failure. Thus the state could act to offer finance or could work as a forum to agree common standards whilst overall seeking to exert minimal control over network development. There is limited need for direct regulation as the market would tend to regulate itself but actions to secure universal service provision may have to occur. Action may also involve ensuring that the fragmented development of networks does not result in sub-optimal solutions. Hence limited pre-competitive interactions may be allowed in areas such as standards. Overall, the priority of government in this scenario is to create the necessary commercial environment in which firms undertaking investment in infrastructure can expect a satisfactory return. Thus, there will be little direct involvement by the state in network provision.

Strategic

These are active policy measures taken within the context of a competitive market situation. This is perhaps the most common form of network development, especially when the system is in transition towards a competitive system. The introduction of competition into the network sectors has encouraged substantial private investment into networks to support all aspect of their development. However, as much as the private sector leads the development of infrastructure, there is still direct state involvement *via* financial support (both direct and indirect) to encourage private sector providers to meet the broader socioeconomic objectives. Thus the nature of the interface between corporate strategy and public policy is based around policy makers seeking to influence market outcomes *via* their actions. These can be at the strategy level where the business and government interact to create a network development programme that fulfils both their needs or in pre-competitive actions *via* social programmes, etc.

This interface between network providers and the state reflects a concern by the latter to ensure that competitive structures do not undermine the broader socioeconomic remit and concerns inherent within the development of infrastructure. Thus finance may be offered from state to private providers to develop networks of broader socio-economic interest. Though policy will tend not to intervene if linked to economic progress. The state will also allow pre-competitive interaction between players and may act as co-ordinating forum for dispersed and devolved actions of the private sector. Furthermore, the state may act as a catalyst for innovation and knowledge creation. As a result, finance for network development may be based on a combination of public and private sector actions. However this does open up the risk of policy failure as the state may end up doing what the private sector would have done anyway or end up promoting products that are simply not desired.

* * *

The form and nature of the creation of transnational network infrastructure is linked into the prevailing paradigm for network development. Under a system of 'control', limited interconnection is promoted to meet the broad objectives of industrial policy. Foreign operators have few, if any rights, to create a uniform network environment. Thus internationalisation takes the form of a series of governmental/incumbent agreements regarding limited integration to meet broad strategic and socio-political obligations. The 'strategic' pattern of development suggests market-led internationalisation with the policy/strategy interface based on governments seeking to influence market forces so that networks internationalise along the lines of their industrial policy. The aim is to use active support to counteract market failure. Under the 'corporate' model of development internationalisation occurs through inter-incumbent agreement to integrate networks in a manner that is compatible with their broad strategic objectives. Policy can seek to influence this process through regulation, etc. Finally the 'liberal' paradigm of network development generates integration and internationalisation through market processes. There is limited pre-competitive or state involvement in areas such as research and development and standards setting.

Each pattern has its own discrete political economy in terms of its interface between states and the markets. These pressures are directly derived from the interface between the market/non-market strategies of incumbents and their interface with the objectives of public policy. These operators have an important strategic and political legacy for the development of the network economy. The liberal perspective of network development requires harmony across all (or most states) with regard to the role of commercial forces and limits the political aspect of strategy through effective regulation based on activism and credible sanctions. A pre-requisite of such actions has to be a divorce between the state and the incumbent, not least through the former selling its stake in the latter. Where there is a strong stake, there is a direct incentive both politically and financially to limit the intensity of competition faced by these entities. This implies that despite internationalising markets there are still pressures to sustain fragmentation.

Conclusion

The aim of this text is to use the framework for network development as a platform for the rest of the book. The intention is to use this paradigm for network development to account for and analyse developments within the development of the European network economy as represented by the development of TENs. Across the transport, energy and telecommunication sectors, analysis will focus on the pattern and trend in network development and on the nature of the strategy policy interface through the paradigm developed above. Through this device, we seek to account for progress towards and impediments in the integration of infrastructure networks across Europe. This is despite a more cogent effort by the European Commission to generate a network of networks across Europe *via* the trans-European network programme. It is to this policy measure that the attention of this text now turns.

References

Allen, D. (1988) 'New telecommunications services – network externalities and critical mass', *Telecommunications Policy*, 12(3): 257–71.

Bartle, I. Müller, M., Sturm, R. and Wilks, S. (2002) *The Regulatory State: Britain and Germany Compared*, London: Anglo-German Foundation for the Study of Industrial Society.

Baumol W.J., Panzer J. and Willig R.D. (1982) *Contestable Markets and the Theory of Industrial Structure*, New York: Harcourt Brace and Jovanovitch.

Bonardi, J.P. (2004) 'Global and Political Strategies in Deregulated Industries: The Asymmetric Behaviors of Former Monopolies', *Strategic Management Journal*, 25(2): 101–21.

Buckley, P. and Casson, M. (1976) *The Future of the Multinational Enterprise*, London: Macmillan.

Davies, H. and Ellis, P. (2000) 'Porter's Competitive Advantage of Nations: Time for the Final Judgement', *Journal of Management Studies*, 37(8): 1189–213.

De Man, A.-P. (2003) *The Network Economy: Strategy, Structure and Management*, Cheltenham: Edward Elgar.

Dunning, J. (1991) *Explaining international production*, London: HarperCollins Academic.

European Roundtable of Industrialists (1990) *Missing Networks In Europe*, Brussels: ERT.

Federico, G. and Forman-Peck, J. (eds) (1994) *European Industrial Policy: The Twentieth Century Experience*, Oxford: Oxford University Press.

Fletcher, R. and Barrett, N. (2001) 'Embeddedness and the Evolution of Global Networks: An Australian Case Study', *Industrial Marketing Management*, 30(7): 561–74.

Gadde, L.-E., Ford, D., Hakansson, H. and Snehota, I. (2003) *Managing Business Relationships*, Chichester: Wiley.

Gulati, R., Nitin, N. and Akbar, Z. (2000) 'Strategic Networks', *Strategic Management Journal*, 21: 203–15.

Jarillo, J. (1993) *Strategic Networks: Creating the Borderless Organization*, Oxford: Butterworth-Heinemann.

Johnson, D. and Turner, C. (1997) *Trans-European Networks: the Political Economy of Integrating Europe's Infrastructure*, Basingstoke: Macmillan.

Kogut, B. (1983) 'Foreign direct investment as a sequential process', in C. Kindelberger and D. Audretsch (eds) *The Multinational Corporation in the 1980s*, Cambridge, MA: MIT Press, pp.35–56.

Lovelock, C. and Yip, G. (1996) 'Developing Global Strategies for Service Businesses', *California Management Review*, 37(3): 64–86.

Mansell, R. (1993) *The new telecommunications: a political economy of network evolution*, Thousand Oaks, CA: Sage Publications.

Matheu, M. and Jeunemaître, A. (2002), *Regulation of Network Utilities: the European Experience*, Oxford: Oxford University Press.

Nijkamp, P., Vleugel, J., Maggi, R. and Masser, I. (eds) (1994) *Missing Transport Networks in Europe*, Aldershot: Avebury.

Porter, M. (1990) *The Competitive Advantage of Nations*, Basingstoke: Macmillan.

Rugman, A. and Verbeke, A. (1993) 'Foreign Subsidiaries and Multinational Strategic Management: An Extension and Correction of Porter's Single Diamond Framework', *Management International Review*, 33(2): 71–84.

Schmid, S., Schurig, A. and Kutschker, M. (2002) 'The MNC as a network – a closer look at intra-organizational flows', in S. Lundan (ed.) *Network Knowledge in International Business*, Cheltenham: Edward Elgar, pp.45–72.

Strange, S. (1994) *States and Markets*, London: Pinter.

Utton, M. (1986) *The Economics of Regulating Industry*, Oxford: Blackwell.

2
A Development Paradigm for Trans-European Networks

The development of trans-European networks (TENs) is central to the EU's integration strategy through underpinning the smooth functioning of its constituent economies. Traditionally, European infrastructure was developed in a piecemeal fashion by nation states (*via* collective provision) to meet their own requirements. TENs, if realised in the form envisaged by the Commission, will represent an important shift in the pattern of network development as their evolution will be more pluralistic and market driven. This chapter sets out the broad European context for the framework of network development identified in the previous chapter by noting and accounting for shifts in the policy/corporate strategy interface that are driving TENs. The objectives are to note the process of regime change within infrastructure development as it shifts from a process of control and to examine the role that supranational forces and their interaction with national political processes play in generating this process of change.

Control and fragmentation: the national development of European infrastructure

That infrastructure has a pivotal and profound importance for social, economic and political development is well understood. Major economic breakthroughs have been accompanied throughout history by improvements in production, communications and logistics. That is not to say that such improvements cause economic transformation but they have often been necessary to it. The logistical developments associated with the expansion of maritime trade, the domestic inland transport systems, railways and the transport revolution linked broad economic change with advancements in the state of national infrastructure (for a review see Nijkamp *et al*, 1994). This has been reinforced by the more recent logistical shifts stimulated by the spread of the Internet.

Historically, the dominant paradigm for the development of infrastructure networks within Europe has been through control (often *via* state ownership) with funding occurring through collective provision. The objectives of this 'traditional paradigm' were to control the development of infrastructure to ensure that it addressed a broad set of socioeconomic and political concerns. This regime was born of a mistrust of the role of market forces and their ability to ensure these broad set of concerns were adequately addressed. Thus state ownership and control dominated along the network value chain (services, infrastructure, etc.) with finance often a mixture of state subsidy and state controlled/determined user charges.

Given the aforementioned nature of government objectives and the (then) limited intensity of global competition, the primary objective of this regime was to meet domestic needs – international needs were of secondary concern. In this era, there was limited cross border functionality and flows (such as those with telecoms and energy). Where such flows were evident, interconnection and integration occurred through intergovernmental and industry-wide (through bodies such as the ITU and UCTPE) agreements. Such agreements had limited functionality and served to reinforce the strategic aims of the respective governments by leaving their control over their respective industry value chain intact. However as a result of this controlled and fragmented approach to infrastructure development, inconsistencies emerged in international infrastructure. In an era of limited international interaction, this mattered little but as the commercial environment changed, this pattern looked anachronistic. It is only relatively recently (from the mid-1980s onwards) that the transnational and even global aspects of infrastructure development began to be stressed as the network economy and its implications rose in prominence.

The programme of economic integration launched under the Treaty of Rome started to challenge the policy framework that sustained international fragmentation. The result was that the control paradigm of network development began to be challenged. This was highlighted most notably by the programme launched in the mid-1980s to complete the Single European Market (SEM). It was evident that if mobility in law was to be met by mobility in practice then a system of control that engendered fragmentation had to be removed and replaced by one that stressed a measure of integration.

Internationalisation under control: an emerging infrastructure policy for the European Union

The beginnings of a supranational challenge to this process of control began rather tamely. In the 1950s and 60s, any supranational action had to take place against a background of sustained control by the respective states of the development of network infrastructure. Initially, action was limited in focus through the formulation of the Common Transport Policy (CTP): the other TENs sectors received some attention but in a more fragmented way. The focus on transport was driven by that fact that it was the only sector (along with agriculture) to get its own title within the Treaty of Rome. The three objectives of the CTP (harmonisation, implementation and competition) represented the beginnings of a challenge (albeit limited) to the prevailing fragmented system based on national control. The CTP, as configured in the Treaty of Rome, meant less focus on national needs, a shift in control from states to markets and a reduction of total control of states over the transport system. The Commission rationalised such challenges on the premise that if infrastructure and networks were to be aligned, then more harmonious conditions for transport had to be developed across the continent. The CTP stressed that states needed to look beyond their own borders and recognise the spillover effects of domestic infrastructure on other states.

In a European Community of limited size, there was greater harmony between states who agreed to co-ordinate national transport systems. Indeed in 1960, in co-operation with member states, the Commission began work on a blueprint for a network of major Community arteries for each mode of transport and on proposals to ensure the co-ordination of investment. These proposals recognised the need for the transport network to evolve with the common market, though states were still unwilling to cede control. Often states' willingness to allow and discuss trans-European networks was often mercenary as such discussions were often linked to their ability to access community funding and to free up internal resources. In a more passive manner, agreements were reached on issues regarding interconnection and interoperability. These were important as they allowed for cross-border flows and for the emergence of a co-operative framework and understanding amongst national governments to reflect emerging interdependencies between states.

In 1966, the Council took an important step towards challenging the totality of control exercised by member states over European transport when it introduced a formal procedure for consultation on infrastructure projects of Community importance. Whilst attempts to reduce elements of control on areas of supranational importance were curtailed, states were required to inform the Commission of projects of Community interest in transport. 'Community interest' at this stage largely referred to transport and trade between member states, or between the Community and third countries. Whilst the Commission recognised a need to go further in terms of challenging control in transnational infrastructure, little was done until 1974 with the publication of new guidelines for the CTP. This and the 1978 Council Decision spoke of the need for enhanced co-ordination and gave the Commission increased consultation rights in infrastructure falling under its remit. From 1978, the European Commission shifted from playing a facilitating role in the consultation process to seeking a more active role for itself in the development of infrastructure with a European dimension (CEC, 1979).

These measures underline that power still rested with the states. All the European Commission could seek to do was create a forum in which the respective states could meet and seek to co-ordinate the transport systems that they each controlled. The supranational bodies had little if any formal power to challenge the mode or method of infrastructure planning. States still persisted with the post-war paradigm of active industrial policy drivers of infrastructure development. Whilst the Commission rationalised a more active role within the Memorandum to the Council on *The role of the Community in the development of transport infrastructure* (CEC, 1979), there was little progress in terms of wresting control from states in areas where overlap was apparent. The major policy measures are summarised in Table 2.1.

Despite attempts to gain a more active role in infrastructure development, there was little direct challenge to the prevailing model of control. In practical terms, the Commission pressed ahead with its infrastructure policy but no further significant breakthroughs occurred until the end of the 1980s. In part, this was driven by the stalling of economic integration in the early 1980s and it was only with the launch of the SEM programme that any momentum was restored. The SEM caused business to consider new opportunities and challenges that gave renewed impetus to the campaign for infrastructure with a Community dimension. The first fears about inadequate infrastructure damaging the SEM were expressed most effectively by the European Round Table of Industrialists (ERT), a think tank comprising the heads of several major European companies (ERT, 1989). Similar concerns

Table 2.1 Pre-single market infrastructure policy

Policy measure	Actions	Nature of challenge to control of states
Treaty of Rome – Articles 74–84 (1957)	Development of Common Transport Policy	Seeks common rules for transports sectors and transport between member states. Seeks to introduce competitive transport system.
Formal Procedure for Consultation on Infrastructure (1966)	Commission involvement on projects of community interest	The Commission gained a right of consultation on specified projects. Role largely facilitatory.
New Guidelines for CTP (1974)	Noted need for trans-European network development	Promote advancement of network of networks. Stress co-ordination.
1978 Council Decision	Expansion of community interest	Expanded scope of consultation of Commission.

were expressed by heads of government at the 1989 Strasbourg Council and were reinforced by the Dublin Council the following year. These concerns took place against the beginnings of a re-examination (at the national level) of the paradigm of infrastructure development based on control.

The emergence of modern TENs: from the Action Plan to the Treaty on European Union

Whilst the SEM led to a re-examination of the importance of pan-European infrastructure, there was recognition that developing the European dimension of networks faced considerable barriers within the prevailing framework of development based on control. The system of control was synonymous with a fragmented system of development that was counterproductive to the aims of objectives of the SEM and possibly (over the longer term) to the needs of national economies. Whilst some states were already beginning to gradually shift the mode of network development, such improvements were piecemeal. The 1990 Action Programme was a landmark document which marked the beginning of the modern TENs programme. The document contained the first systematic elaboration of the principles underlying TENs and was symptomatic of the supranational driven shift in the form and nature of network development. The Action Plan highlighted the complementarity of TENs and the SEM in five major areas (CEC, 1990).

1. The volume effect: infrastructure must be able to respond to and accommodate anticipated increases in intra-EU trade resulting from the SEM and greater economic interdependence generally;
2. The interoperability requirement: mobility requires the networks to be physically compatible in terms of services offered;
3. The dimension effect: infrastructure planning should take not only the national but also the European dimension into account;
4. The quality requirement: efforts need to be made to ensure that the development of infrastructure, and thus the quality of service provided, does not vary over space and lead to differences in regional development;
5. The cohesion effect: infrastructure development at the EU level should seek to promote the more even development of the European economy.

These objectives highlighted the need for a change in the mindset of infrastructure development, stressing that both market-driven (based on the need to be responsive to anticipated traffic flows) and transnational concerns must be considered in network evolution. Whilst this did not necessarily challenge the system of control, it did have implications for the domestic collective provision of infrastructure as the existing processes created free riders and limited the ability of domestic industrial policy to isolate and maximise the national impact of its actions. Thus in identifying TENs as a policy priority, the EU sanctioned a policy change to promote the market-led development and private provision of those networks with a pan-European dimension. This represented a potential sea change in the pattern of infrastructure development within Europe. However, it left the prevailing national models intact.

To promote the market-led development of pan-national infrastructure, the Action Programme proposed three courses of action.

1. The identification of priority projects that possess a clear European dimension, in that they aid the functioning of the SEM, support moves towards EMU and support trade on a broader pan-European basis;

2. The promotion of horizontal measures to ensure the emergence of a coherent set of networks through the development of masterplans, regulations, standards to foster interoperability, feasibility studies and the creation of a databank to collate and evaluate requirements for, and barriers to, the construction of TENs;
3. Financial measures to attract private support for the priority projects. The preference for private finance implies that projects must be profitable, possibly with the support of public financial intervention. However, in light of the scale of the expenditure required, private finance will have to play a bigger part in large-scale infrastructure projects than hitherto.

These measures challenged the prevailing paradigm of control by stressing that not only should the external dimension of national infrastructure be considered but also that both EU institutions and the private sector should play a more prominent role in their development. This latter point is especially important in that the pressure for markets to lead and finance the development of infrastructure was the beginning of a paradigm shift in network development; a process that was closely tied into the liberalisation process. This trend was cemented with inclusion of trans-European networks in the Treaty on European Union.

The TENs remit in the Treaty on European Union stresses many themes but it is evident that if economic union is to be achieved, TENs will be a key component of that process. TENs are not only a complement to mobility and to the objectives of the SEM but they must also be viewed as a means to enhance growth across all parts of the EU and promote more even economic development. In short, TENs are a step forward in the process of economic unification which is a key precursor of stable economic and monetary union (EMU). Moves towards sustainable EMU rely upon the economic convergence of the component parts of the EU (see below). In this context, TENs are regarded as a market-based solution to regional disparities by reducing, in temporal terms, the distance between core and periphery. In emphasising the importance of infrastructure in economic integration, the Commission sought to add political impetus to the process of reform of network development at the nation state level.

The Treaty on European Union, and its treatment of TENs, was important as it not only committed all EU states to the deregulation of network services but it also shifted the focus of development from a network of networks that has traditionally dominated network development towards a more integrated approach. This directly challenges the traditional paradigm of network development as activism, monopoly and control starts to be undermined by rising interdependence and through a push for more market-driven evolution.

The Treaty was a green light to the Commission to press ahead with the development of trans-European networks. By the beginning of the 1990s, against a backdrop of a Europe-wide recession, TENs moved to centre stage of the EU's competitiveness programme and became its big idea for dealing with many of its economic problems. This was aided by an environment of restricted member state financial freedom and an emerging liberal hegemony. These trends were making states more aware of the need for the internationalisation of infrastructure and of a need to shift the prevailing paradigm of network development and to look towards more diverse forms of provision and finance (CEC, 1993). By 1995, work to further the TENs goals of the Treaty on European Union was well underway. Guidelines for the three major networks were presented by the Commission in 1994 (CEC, 1993). Simultaneously, groups of personal representatives of heads of state, in the form of the Bangemann Group (telecommunications) and the Christophersen Group (transport and energy), were investigating ways to promote these networks.

The shifting paradigm of network development

Whilst it can be argued that TENs are essentially a technocratic issue, they also represent an important impetus to the process of integration. Developing infrastructure that meets purely national needs is not only anachronistic but increasingly self-defeating. In particular, the degree of interaction between states is such that it is increasingly unlikely that any large-scale infrastructure project will have purely national effects. In political terms, these efforts are a classic example of neo-functionalist spillover which offers a coherent rationale for a direct challenge to the process of national control over infrastructure especially where there is a clear international dimension. It underlines that a broader set of concerns need to be considered that may not be justified under a system based on control. Under a system of control, the state may find it hard to legitimise using domestic finance to develop infrastructure where there may be direct (and indirect) non-paying beneficiaries.

As an exercise in negative integration (see Tinbergen, 1954), the TENs programme reflects the desire of states to overcome the impediments to interaction created by a fragmented system. As the rationale for the sustenance of a fragmented system has been undermined by the internationalisation of business, so it has become evident that infrastructure is an important (albeit passive) unifying tool of political and economic integration. These changes increase the degree of Europeanisation of individual economies and render them increasingly dependent upon conditions within the European economy as a whole. These networks act as a forum for interaction between economic, political and social actors to which the organisation needs to respond. Integration is increasingly occurring as a practical response to everyday economics. As Wallace (1990) points out, the dominance of informal, market-based integration over formal, intergovernmental integration has had greater effect in pushing the process forward, that is, integration as a practical necessity not as a deliberate political decision (for an examination of integration issues see Robson, 1998).

TENs once established will, the Commission hopes, create a sense of dynamism to push forward the process of integration by 'informal' means. It is in this instance that the interdependence between negative and positive integration becomes apparent (see Tinbergen, 1954): the benefits of the former are only really effective if they are supported by the latter. That is to say, the establishment of a common resource to meet the needs of economic actors requires a common system of management. The nature of TENs as a common resource requires that private interest should not take precedence over broader public interest. Such shifts linked with the emergent network economy in Europe are evident within the following key underpinning themes of the TENs programmes.

- Interconnection is the physical linking of separate national networks to form a single, all-embracing European network. Network configuration within individual states tends towards a core-periphery model. Consequently, infrastructure tends to be less developed in border and frontier regions. The need for interconnection of national networks also pinpoints those areas where there are apparent bottlenecks due to the underdevelopment of networks. The result of such changes is that the national network becomes a sub-network of a larger whole. In many senses, what is actually created is a 'virtual network' as the European network is a series of interconnected national networks;
- Interoperability requires the technical compatibility of linked networks. If the interconnection of networks is to be beneficial, services offered by networks in one member state should be readily

transferable and operable on equivalent networks in other member states. Traditionally national networks have developed according to different specifications and technical regulations. Developing common standards for operation is central to realising the benefits from TENs. Interoperability has to be extended to technical equipment and facilities as well as services. This requires a re-regulation of these sectors to ensure that services are developed in a harmonised fashion across the area. The development of interoperability offers greater scope for economies of scale and, as a consequence, lower costs in terms of network services;
- Access is crucial to the success of TENs: not only must networks be physically integrated and their services compatible but access to them must also be free and open. In effect, the development of TENs should not include the possibility of discrimination in terms of use or actual exclusion from the network. Access is central to the efficient use of the existing networks and will be important in ensuring that private finance to these networks is feasible, especially if revenue is to be based upon user charges. All users should be treated equally in terms of access and no preferential treatment should be given;
- Open and competitive markets provide the framework for the establishment of these networks. This has vast implications for infrastructure development. First, infrastructure should only develop where there is an evident need for that infrastructure. Secondly, infrastructure development should complement the broader market changes that are happening on the European continent. Thirdly, because of the market origin of these networks, it should ideally be through this mechanism that they are realised. That is, the market will indicate what pieces of infrastructure are necessary for the functioning of the economy thereby giving private investors an incentive to provide it. This is important: TENs are not simply a measure for short-term political need but have an important function to play in the even development of the SEM.

The informal integration facilitated by the development of TENs is likely to exert greater pressure upon the more formal means of integration. The development of TENs and the exploitation of their effects relies, in the short term, upon enhancing and promoting the more efficient and effective use of existing networks by assisting their integration and interoperability and by ensuring that access is non-discriminatory. These informal pressures generate pressure on the form and pattern of network development. The emergent network economy has to be responsive to user needs if it is going to aid the performance of the European economy. Existing power structures supported by a fragmented network system offer no rationale as to why they should be responsive to such needs and why transnationality should be considered.

These themes represent a sea change in the proposed pattern of network development especially at the international level. These patterns suggest that the process should be increasingly dependant upon the market for both delivery and finance. This represents a shift away from the notion of infrastructure development to meet national needs and paid for by state coffers. However, this should not be represented as a direct challenge to the states. As mentioned, tighter budgets and the growing realisation of the impact of economic integration were already reducing parochialism and promoting an examination of alternative means to fund the development of infrastructure. This was symptomatic of a shift away from active industrial policies (see below) as many states were stepping back from extensive involvement in the domestic economy through progressive deregulation. These themes merely added impetus to the process.

The network utilities were initially excluded from the development of the SEM for the reason that it was highly unlikely that any political agreement would be forthcoming among member states. Thus a separate set of initiatives were launched centring on the liberalisation of energy and telecoms. Across both sectors, the Commission's strategy was one of an incremental encroachment of market forces to gradually expose the natural monopoly core. However across both of these sectors, the integrative approach to the introduction of market forces was resisted by incumbents through both market and non-market strategies.

Thus, despite states moving away from control and ownership towards a market-based paradigm, the actions of incumbents suggest that they did not share the same zeal to relinquish their dominant position. Thus despite a commitment to the move towards a liberal regime, the shift away from control pushed the development in infrastructure in a number of different directions. In some cases, incumbents resisted market erosion and sector-based regulation became common within a corporate pattern of development. In others, the state only half-heartedly accepted its commitments allowing limited competition within a framework of sustained public sector involvement (McGowan, 1993). This gradual process of regime shift is highlighted within Table 2.2.

Table 2.2 Regime shift within European Union infrastructure provision

Era	Member state level regime	EU involvement in network infrastructure	Market-based actions	EU Network System
1950s–mid-1980s	Control	Consultation	Control-based integration via common policy	Fragmentation and control
Mid-1980s–mid-1990s	Strategic	Co-ordination, Consultation	SEM-driven integration, market-based, limited to specific sectors	Interaction and incumbency driven
Mid-1990s – current	Strategic/Liberal	Co-ordination, Consultation, Facilitation	Market-driven integration extended to all aspects of infrastructure	Pluralistic network development and integration

Industrial policy and the competitive environment

At both the EU and the nation state level, the movement of TENs up the policy agenda coincided with the shift towards a more general industrial policy, moving away from the traditional sectoral and active approach. Within this context, policy is centred upon creating the right conditions in which the market can operate effectively and to the advantage of the EU's economy. This passive role stresses economic stability, freer markets and instruments that push the market to behave in the desired way. In short, market realities are to dictate industrial policy reflecting the Porterian perspective that nothing is better for competitiveness than competition itself (Porter, 1990).

At the EU level, industrial policy consists of a series of measures linked to the completion of the SEM which, if functioning effectively, will provide the necessary impetus to promote structural change. By and large, EU action centres upon alleviating areas of market failure. As a consequence, policy is increasingly geared to enabling European industry to respond flexibly to signals given by the global market. This shift in policy underlines that, ultimately, structural change is up to economic operators and no one else. The new themes in industrial policy are designed to be compatible with the lurch towards subsidiarity, stressing that competitiveness is ultimately up to member states. The EU merely has a co-ordinating role through taking horizontal measures *via* a series of common policies. In part, this is a realisation that the degree of change needed and the expense involved is beyond the capabilities of the public sector. Therefore, this policy for industrial competitiveness, which relies on market forces to be the advocate of change, may not necessarily be the most desired but may be the only practical strategy.

This is borne out by the strategy towards TENs which has focused on a co-ordinating role between the infrastructure developers, consideration of finance issues and ensuring that these networks develop in line with its economic priorities. The reorganisation of production fostered by TENs is seen as a way of promoting industrial competitiveness and of responding to the challenges represented by the new generalist policy (CEC, 1998). The strategy draws a clear distinction between who does what within the context of the completion of TENs. As stated above, TENs activism is curtailed due to differences between required and available resources. Indeed excessive EU (or for that matter state) activism may be incompatible with the development of TENs in the desired fashion as market and political priorities may conflict. The greatest contribution that the Community can make, therefore, is to establish the conditions under which the funding for these networks can be forthcoming. This rests on the following three interrelated elements (CEC, 2002):

- Co-operation between undertakings: the Community has acted as a forum for co-operation between enterprises to develop networks. For example in telecommunications a regulatory framework that defines where co-operation can and cannot occur has been established. Such co-operation in the form of public/private partnerships lies at the heart of the development of the other networks. However, the Commission has been slower in offering regulatory guidance in the energy and transport sectors;
- Co-ordination between operators: to ensure networks develop in a harmonious and compatible fashion across the EU, forums have been developed to oversee the standardised introduction of these networks;
- Re-regulation: if commercial operators are to become involved in the development of TENs, then opportunities must exist to enable them to profit from their investment. This relies upon allowing them greater discretion over what they can do as well as clearer guidance over what they should do and what they must do as network operators. In telecommunications, the gradual process of liberalisation has fostered greater investment in infrastructure to meet increased competition.

Whilst competition is seen as the way forward, this often has to be developed within a co-operative framework given the issues involved.

However, the need to co-operate must be tempered in such a way that it does not exclude broad entry. Creating TENs based upon monopolies will offer little advantage to the EU. The traditional market failure perspective on network development has broken down and there are clearly areas in which market forces can now operate. It is evident that in targeting TENs projects; there is clearly faith in both the energy and telecoms sectors that private finance can deliver all aspects of European infrastructure. There is evidently more concern (as Chapter 3 indicates) in the transport sector. However, the ability of this policy shift to stimulate regime change depends on the willingness of member states to embrace its implications.

The TENs programme and the Lisbon Agenda

Whilst enlargement and EMU have seen TENs slip down the agenda as the Commission's big idea, they remain key to the success of both processes. However, policy statements have veered away from the declaration of grand projects and stressed a more pragmatic policy based on smaller scale measures. Policy towards networks have been driven more by a liberalisation agenda than by targeting specific infrastructure *per se*. The Lisbon Agenda once again moved TENs up the EU's policy agenda linking broader socioeconomic change to the platforms provided by network infrastructure (CEC, 2003a). The absence of TENs was not only hindering the development of an integrated market but was also imposing substantial costs upon business. This problem was compounded by the fact that the need for a widespread regime change in terms of network development was not accepted by all. Indeed the move from control towards liberalisation has been more protracted for some states than for others with the result that competition has developed unevenly notably in terms of the cost and quality of service offering.

In the aftermath of the Lisbon Agenda, a ten year programme was launched which identified the need for a quick start programme (see below) for the development of TENs where the problems were more administrative than financial. These are ready to launch projects which are relatively mature, have a trans-frontier dimension impact on growth and benefit the environment. These are in areas such as technology in transport and aiding the spread of broadband and are designed to have an immediate short-term impact upon improving the quality of infrastructure. With limited potential for large scale public sector involvement, any relaunch of TENs required a further shift towards market-driven processes. Thus, there is a provision for finance from the EU to consider feasibility of projects to be market driven where network development is linked to enhanced freedoms from the provision of network services. However these measures have to be set within the context of the Services of General Interest directive. This directive seeks to ensure that the liberalisation of services does not work to counteract socioeconomic cohesion and is a means through which some of the conventional themes of infrastructure development under the traditional method of control are kept. This directive is a legacy of the system of control and of the conventional mistrust of markets to deliver universality in network services. In many ways, this reflects the fact that moves towards a more market-focused industrial policy is still set within the context of the European social model of inclusiveness and cohesion.

The more pragmatic approach to the development of TENs' priorities stresses less ambitious short-term aims for networks to contribute to economic development and integration through, for example, encouraging more efficient usage, etc. These projects are based on low capital investment, low sunk costs in areas where private financing will be easier to attract than in the big, highly ambitious projects previously targeted. At the EU level, the following areas are prioritised:

- The revitalisation of railways through broad re-regulation;
- Adoption of the Eurovignette to allow proper charging of infrastructure;
- The development of the single European sky through air traffic reform;
- Support for Galileo to aid advanced navigation systems;
- Progress with energy liberalisation;
- Promotion of energy efficiency;
- Increased use of renewable energy.

At the member state level, the aim is to seek renewed commitment to existing liberalisation measures, to generate investment for projects of 'European interest', to develop intelligent transport management systems and to reform planning to stimulate a European dimension within projects.

As a result of this more pragmatic approach, the TENs programme has become more focused, offering concerted support whilst being proactive in developing international co-operation through the appointment of project co-ordinators to implement specified priority schemes (see below). However any direct support in terms of finance is merely incidental to the overall levels required. As of 2006, the sources of finance from the EU are:

- The TENs budget line (up to €700 million per annum up to 10 per cent of cost);

- Structural funds (of up to €60.9 billion on infrastructure and related development between 2000 and 2006);
- The sixth framework programme is developing €16.5 billion for RTD.

In addition, there is funding from the European Investment Bank (EIB) through the TENs investment facility, research and development and support for innovation and risk capital.

It was estimated in 2003 that the financial gap in the priority projects identified by the Commission was around €235 billion (CEC, 2003a). In addition, infrastructure investment undertaken by member states has fallen to 1 per cent of GDP (2003b) down from 1.5 per cent in the 1980s. This spending is a fraction of what is needed to ensure the effective development of TENs with the result that the objective of getting key pieces of infrastructure in place by 2010 will be missed. It is estimated that at least an extra two decades will need to be added to their realisation time (CEC, 2003b). The problem is not one of funds but of financiers unwilling to offer finance. This has been compounded by the absence of common regulations and agreed tariffs between states which are needed to ensure there is a route through which the private public partnership will allow the former to enjoy greater commercial return and certainty. In this context, a more liberal regime in network services is the key as freedom in this aspect of the network is expected to feed through into infrastructure development (regulations allowing). However moving from investment in services to infrastructure does represent a significant upgrade in the risk faced by private concerns. As a result, new sources of finance are being sought such as the provision of third party equity or quasi equity (especially for venture capital), securitisation (to increase pool of resources from financial markets by reducing liquidity constraints in areas of important projects) and other mechanisms to spread risk. To this end, the EU is preparing a Guarantee instrument to reduce risks in specific areas (such as the construction phases of projects).

Internationalisation and the shifting paradigm of infrastructure development in the EU

Overall these processes highlight three major trends that are driving regime change in Europe. First has been the change in the policy environment of the member state which has increased liberal policies in network development. Second have been attempts by the EU to shift the regulatory regime across the network sectors. This has been an incremental approach starting with services gradually moving towards the provision of infrastructure itself. Finally integration is internationalising infrastructure with the result that there is increased credibility to the option of the EU as a co-ordinating and facilitating body. These measures highlight that the coalescence of market integration, re-regulation and an emergent need for co-ordination are driving a shift in the paradigm of network development.

The emergence of TENs as a network of national networks is being undermined by a combination of technological and commercial change as dedicated transnational infrastructures emerge funded by commercial forces (Exel *et al*, 2001). Leaving the development of international networks to the states has limited interconnection and competition as it left national systems open to regulatory capture *via* the non-market strategies of incumbents. Helm (2001) argues that this legitimises the development of European regulatory bodies to oversee the development of international infrastructures. Without an overarching perspective, national policies that continue to dominate infrastructure development will sustain parochialism despite the overlap between respective national systems. The Federal Trust has suggested just such a body in the form of a 'European Infrastructure Agency' (Federal Trust, 1996). This body would be mandated to develop the pan-European dimension of infrastructure and to address market failures in their development as well as seeking to establish new sources of finance. The Federal Trust stressed that the body would have to be independent of existing organisational structures and set within the context of the market-driven development of TENs.

However despite being a policy designed to recognise the evolving interdependence of infrastructure, the TENs programme sustains a strong national dimension as many of the priorities do little more than sanction national plans and lack a general global view. Many have remarked on the need for strong action to create a regime that stresses the market on an international scale but few concerted actions have emerged, especially within transport. Within the context of subsidiarity, Turro (1999) suggests that any pattern of infrastructure development has to be based on:

- States leading the development of TENs nationally with the Community guiding actions;
- Integrating TENs within other policies at both the national and supranational level;
- Integration of networks both across and within sectors;
- The ability of TENs to offer door-to-door benefits.

Sichelschmidt (1999) highlights that the development of the TENs project depends upon information provided by member states. With limited and biased information, the policy will be both prejudiced and subject to the problems of moral hazard. Thus member states seek to attract money from TENs not because of the integrative aspects but because of the desire to access Community funds (Wink, 1996). This is to be reinforced by the desire of states to get their own peripheral regions better infrastructure which has more local than international benefits. In addition, the focus on the support for rail within the priority projects reflects a member state priority and offers support for projects that were often already funded. Thus the TENs funding is a windfall. This means that many of the jobs created by the TENs programme would have been created anyway and that any increases in employment may have to be offset by losses elsewhere (as transport switches mode for example). In addition, there is no guarantee that improved access to peripheral regions will promote decentralisation. There is the possibility that TENs could have exactly the opposite effect (Vickerman, 1994, 1995) as better infrastructure improves access in both directions allowing firms in central areas to exploit (rather than invest in) these markets.

Despite the move towards introducing market forces, it is evident that there are still market failures across the network sectors. The desire of incumbents to protect their position as well as environmental, social and political concerns means that a complete *laissez-faire* approach to the development of networks in Europe is unlikely. There has been unease at the ability of market forces to fully develop the necessary infrastructures for Europe. The Californian energy crisis, the problems with the UK rail system, the Channel Tunnel and the roll-out of advanced communications networks all highlight problems in the use of market forces in network development. There is also unease about the reliance on overseas suppliers for core inputs and the rampant overblown market for 3G licences. As later chapters show, in some sectors infrastructure has been over-provided, in others it has been under-provided by market forces.

According to Helm (2001) the problem in shifting to market-driven development is that – whilst it allows the utilisation of private sector balance sheets – private investment in infrastructure offered a higher risk premium to markets than that undertaken by the state. In some cases, the trend towards instability in network sectors has driven some states towards a more corporatist model as it encourages co-operation. In Germany and France, there has been a belief that too much competition can be detrimental if it leads to unstable market structures and an eventual under provision of service. In other areas, regulators have allowed consolidation where such actions would not seek a drastic reduction in competition. In part, this has been aided by the segmentation on a vertical and horizontal level of the marketplace. However this consolidation risks shifting the emergence of TENs away from a competitive framework to one based more on strategic concerns. This consolidation has shifted the policy regime. According to Helm (2001) this was the result of a policy of benign neglect by the European Commission which has tended to treat the national market as the relevant market in the assessment of dominance and as consequence ignoring transnational consolidations. The result is that a truly competitive market is now likely to be elusive.

Within and across the network sectors, the pace of liberalisation has varied. The more obviously competitive segments have shifted towards plural structures whereas the more natural monopoly element has retained a degree of corporate dominance and state control. This reflects what Helm calls the 'onion' approach to network development based on a gradual increase in competition which leaves all but the core natural monopoly element exposed. Though, as the previous chapter highlighted, theoretical developments have begun to undermine traditional natural monopoly arguments for limiting their exposure to competitive forces. However these changes have been evolutionary with the result that these shifts towards competitive markets and more passive industrial policies have not occurred at a uniform rate. The result has been unevenness in the development of markets as the state-incumbent interface tends to be stronger in some states than in others. Furthermore, as highlighted above, the vertical and horizontal disaggregation and segmentation of the sectors (and the micro-segments within them) means that these effects are not felt harmoniously across the industry value chain. This highlights that the effect of the regime shift will not be felt symmetrically both across and within sectors. However even in the most reluctant of states there has been gradual acceptance that new ways need to be found to develop infrastructure and this implies greater freedom of entry as a means of generating the traffic needed to justify the levels of investment required.

In part, these problems have been driven by the fact that the move towards a more liberal regime has replaced control with regulation.

Thus market structures generally remain intact, especially where regulators have concerns above and beyond those linked purely to the market. Such actions can work to limit the intensity of competition and delay its onset. These special regulatory regimes – whilst ultimately seen as temporary – do allow control to be exercised.

Conclusion

Against the background of a number of shifting environmental contingencies (such as more mature economic integration and a global shift towards liberal forms of economic management), there is a gradual shift in the form and nature of network development within Europe. However this shift has been gradual as member states sought to retain strong elements of national influence over network development. Whilst there has been a long history of supranational involvement in the development of infrastructure, it is only really since the beginning of the 1990s that it has really started to gain any momentum. However a strategy for the development of a set of TENs needs the agreement of the member states at a number of different levels. First, the Commission needs agreement (and willingness) of states to liberalise the network sectors. Second, there needs to be the necessary administrative structure that allows private finance to be channelled into infrastructure. Third, there needs to be support at the member state level for the Commission to take a more prominent role in the development of infrastructure. Thus TENs rely upon regime shift at the supranational level to be matched and implemented at the national level. This requires a reduction in the national mindset, a reduction in the need for control and an increased faith by governments in markets to deliver socially optimal outcomes across.

Across the EU, there has been a gradual shift towards more liberal regimes. As the following chapters will show across the respective network sectors, this shift has been uneven as well as protracted. Pressure from the EU to push a liberal agenda for the development of TENs has often met with a lukewarm response from the member states. Whilst control tends to be rare (and it can be argued that this hegemonic position has broken down), the shift towards a liberal regime is by no means universal and indeed there may be elements of consolidation pushing the sector towards more corporate system of management. With these concerns in mind, the rest of the book looks at individual sectors in more depth, looking at the process and pace of regime shift within each.

References

Commission of the European Communities (CEC) (1979) *The role of the Community in the development of transport infrastructure*, adopted 7 November 1979, Bulletin Supplement 8/79.

Commission of the European Communities (CEC) (1990) *Towards Trans-European Networks for a Community Action Programme*, COM(90) 585, Brussels.

Commission of the European Communities (CEC) (1993) *Community guidelines on trans-European energy networks and Proposal for an EP and Council decision laying down a series of guidelines on trans-European energy networks*, COM(93) 685, Brussels.

Commission of the European Communities (CEC) (1998) *The competitiveness of European enterprises in the face of globalisation – how can it be encouraged*, COM(98) 718, Brussels.

Commission of the European Communities (CEC) (2002) *Industrial Policy in an Enlarged Europe*, COM(2002) 714, Brussels.

Commission of the European Communities (CEC) (2003a) *Some Key Issues in Europe's Competitiveness: Towards an Integrated Approach*, COM(2003) 704, Brussels.

Commission of the European Communities (CEC) (2003b) *Building Trans-European Networks*, Memorandum, Brussels.

European Round Table of Industrialists (ERT) (1989) *Missing Networks: A European Challenge*, Brussels: ERT.

Exel J., Rienstra S., Gommers, M., Pearman A. and Tsamboulas, D. (2001) 'EU involvement in TEN development: network effects and European value added', *Transport Policy*, 9(4): 299–311.

Federal Trust (1996) *Private Partnerships and Public Networks in Europe*, London: Sweet and Maxwell.

Helm D. (2001) 'The Assessment: European Networks – Competition, Interconnection and Regulation', *Oxford Review of Economic Policy*, 17(3): 297–312.

McGowan, F. (1993) 'Trans-European Networks: Utilities as Infrastructures', *Utilities Policy*, 3(3): 179–86.

Nijkamp, P., Vleugel, J., Maggi, R. and Masser, I. (1994) *Missing Transport Networks in Europe*, Aldershot: Avebury.

Porter, M. (1990) *The Competitive Advantage of Nations*, New York: Free Press.

Robson, P. (1998) *The economics of international integration*, 4th Edition, London: Routledge.

Sichelschmidt, H. (1999) 'The EU programme "trans-European networks" – a critical assessment', *Transport Policy*, 6(3): 169–81.

Tinbergen, J. (1954) *International Economic Integration*, Amsterdam: North Holland.

Turro, M. (1999) *Going Trans-European: Planning and Financing Transport Networks for Europe*, Amsterdam: Pergamon.

Vickerman, R.W. (1994) 'Transport infrastructure and region building in the European Community', *Journal of Common Market Studies*, 32(1): 1–24.

Vickerman, R.W. (1995) 'Location, accessibility and regional development: the appraisal of trans-European networks', *Transport Policy*, 2(4): 225–34.

Wallace, W. (ed.) (1990) *The Dynamics of European Integration*, London: Pinter.

Wink, R. (1996) 'Transeuropäische Verkehrsnetze: Für Wachstum oder für regionale Umverteilung?' *Wirtschaftsdienst*, 76(6): 301–8.

3
Transport TENs

Transport represents the circulatory system of the economy and without it economic activity would atrophy. An efficient, effective transport infrastructure is therefore an essential factor in a nation's competitiveness. Alongside these economic concerns, the development and configuration of transport infrastructure has vital social and political roles. Thus the configuration of transport networks (including road, rails, inland waterways and the location of ports and airports) helps shape economic activity and determines how and where we live and work. Given its strategic importance, the state has long maintained a major role in controlling, planning and financing transport networks.

The significance of having the appropriate quantity and quality of transport infrastructure has been highlighted by the emerging mobility gap in Europe. As a result of contemporary transport trends (see below), the demand for transport services is growing more rapidly than the existing transport networks can cope with. The resulting congestion and bottlenecks are imposing additional costs on business and pose a serious threat to the competitiveness of European business. Problems are also occurring because of inadequate connections between Europe's peripheral regions and the rest of the European trading area.

Across the constituent transport modes and their components (links, nodes, etc.), the state maintains a strong influence and often outright control. However, that is not to underplay the role that the private sector has played across modes and network components throughout Europe. In rail for example, the state's involvement has varied considerably across the continent whilst civil aviation developed under the umbrella of the state and shipping has had a more sustained record of private sector involvement. The political and economic fragmentation of the European continent as well as the complexity of transport and its network components has meant that – unlike other networks – establishing a common pattern of development can be difficult.

The economic importance of transport infrastructure is based on the dual roles that it performs:

- First, the transport sector is economically significant in its own right. In 2001, almost one million enterprises within the European Union were classified as 'transport services'. Altogether they generated €308 billion of value-added in the EU-25 countries, that is 12.7 per cent of value-added in the non-financial services sector and employed 7.4 million people, or 10.9 per cent of those in non-financial services (Eurostat, 2004: 323);
- Secondly, and most importantly, transport infrastructure supports and facilitates general economic activity and commercial transactions. This is the case both within and between nation states. As this chapter indicates, transport networks play a crucial role in European market integration by ensuring physical access to all parts of the European economic space. It is also central to promoting the industrial and corporate restructuring that is envisaged as a consequence of the process of market integration.

The transport needs of business are thus driven, in part, by the need to maximise the benefits from market integration in Europe. As such, transnational transport networks have a role to play in terms of keeping the costs of European business down and of facilitating the logistics revolution outlined below. Moreover, in the same way as many of the transport TENs initiatives challenge (see below) member states to connect their transport networks to overcome the missing links between them, so transport TENs can encourage member states to do the same with neighbouring countries. Measures to improve traffic circulation in the Mediterranean region and South East Europe fall into this category. Other transport policy initiatives involving third countries, such as the EU-US talks on Open Skies, have significant implications for Europe's transport infrastructure and the access granted to it.

The task of developing integrated transport networks is far from easy. Transport TENs encounter all the problems of developing infrastructure with the addition of several specific cross-border challenges. First, transport infrastructure incorporates a variety of modes – road rail, maritime transport (both short sea and long distance shipping), inland waterways and civil aviation. All too frequently, network planning takes place in modal isolation with too little attention paid to intermodal issues. Very often, it is the ease of transfer between different transport modes, both for passengers and freight, that can transform the efficiency of a network and reduce congestion. The 2001 Transport White Paper (CEC, 2001) aims to address the modal balance in a more general way by reversing the trend of the declining share of rail transport in overall European traffic with the help of the transport TENs initiative. Related measures, such as ensuring that the price paid for infrastructure usage directly reflects the costs incurred as a result of that usage, including the costs to society, are an important part of efforts to address modal imbalance.

Secondly, investment in transport infrastructure is the costliest of all infrastructure investment. Moreover, large-scale transport infrastructure investments are notoriously vulnerable to significant cost overruns. The interval between the beginning of the planning stage and the time that a project is complete can run into decades for large-scale transport projects, leading to a very long period before a private company can expect to get a return on its original investment. This return is also uncertain because of the potential cost overruns and risk of traffic forecasts. In short, the size of the investment needed can be overly burdensome for the public sector, particularly during times of strict budget discipline, and insufficiently attractive or too risky for the private sector. That is not to say that public-private partnerships, as hoped for by the original architects of transport TENs, cannot be a feature of TENs in this sector but they have perhaps not been as forthcoming as originally hoped.

Thirdly, in relation to cross-border infrastructure, the EU is limited in what it can achieve. According to the Maastricht Treaty, the EU's role is limited. This means that a top-down approach to cross-border infrastructure design (that is, design of infrastructure that is carried out from an entirely European perspective and is blind to national borders) is not a practical option. Rather, transport planning has always been and remains, a national activity carried out by national planners according to national policies and priorities and constrained by national borders. What may be the best solution from an economic perspective may not necessarily, and increasingly so with the pressures of European integration and globalisation, coincide with these borders. In short, infrastructure within Europe is designed within a national framework and generally does not take the international dimension into account. For example, the appraisal of a national project will be carried out by the national planning authorities with a view to the costs and benefits of the nation into account. But national infrastructure can play an important role in relieving congestion or removing bottlenecks for trans-European transport, particularly if it is located in a transit country, and a full economic appraisal will take this into account. In some cases, the European dimension may be what transforms a project from being economically non-viable or marginally viable into one that is clearly economically viable. However, taxpayers in one member state are, as things stand, unlikely to be prepared to finance infrastructure where significant benefits accrue to nationals of other states.

Fourthly, the national nature of transport planning means that there is a multiplicity of planning operations and planning instruments, making it difficult in practice to plan projects that cross borders, even where there is the political will and commitment to do so. This lack of planning and authorisation harmonisation can be a problem for transport but it is not such a great obstacle for transport TENs as it is for energy TENs (see Chapter 5). Nevertheless, it can lead to delays or even duplication of provision.

The emerging mobility gap and European transport trends

As the growth of intra-EU trade has intensified over the years, particularly since the launch of the Single European Market, and the physical expansion of the EU following each enlargement, so the demands placed on Europe's transport system to support and foster this growing market interaction between the constituent states of the European Union have correspondingly increased. Indeed, the increased commercial interaction between EU states has placed a greater emphasis than ever before specifically on the cross-border aspects of Europe's transport networks. As such, there is an intimate relationship between these transport trends and the TENs initiative. In a world in which market forces were dominant, infrastructure challenges brought to light by the mobility gap would be met by private sector provision. In practice, although the market currently plays a bigger role in transport infrastructure provision than it has done for many years, non-market socioeconomic objectives and problems with financing linked to the size and the commercial risks involved in transport infrastructure continue to inhibit a comprehensive shift from control to a more liberal approach to the development of these networks.

In generic terms, the mobility gap which is driving the need for transport TENs derives from the mismatch between the quantity and quality of transport services demanded and the quantity and quality of transport services supplied. More specifically, the mobility gap manifests itself in a variety of forms, depending on the circumstances surrounding a particular piece of infrastructure. For example, peripheral regions, which may suffer by their distance from the economic core, experience the mobility gap in terms of the unevenness of network development whereas the core regions labour under the burden of congestion. As the analysis below shows, a common transport infrastructure problem is the over-development of road transport at the expense of other transport modes with negative results in terms of congestion and a range of other environmental costs. One response to this is to improve links not only between networks of the same transport modes in different countries but also between different transport modes. Intermodalism not only reduces congestion but makes the less popular modes more viable. Moreover, freedom to provide services in various modes and rights of access to different networks should also help overcome some of the problems posed by the transport trends outlined below and has been promoted at European level (see subsequent sections) as part of the strategy to free up service provision.

The development of transport TENs is taking place within the context of this mobility gap, which in turn is shaped by long-term transport trends. According to the European Commission, personal mobility has more than doubled from 17 km a day in 1970 to 38 km in the late 1990s (CEC, 2001). Moreover, this mobility is increasingly seen as a right, making it difficult for policy makers to introduce measures to reduce congestion or to enforce the polluter pays principle which would require individuals to pay more for their motoring, for example.

Table 3.1 demonstrates that since 1970 road freight has grown much more rapidly than rail freight. Over a period of over three decades, road freight has consistently grown at least two to three percentage points faster than rail freight (see Table 3.1) and has resulted in a greatly increased share for road freight in overall freight transport (see Graph 3.1). In the passenger sector rail performance, although exhibiting steadier growth than freight, was also well below that of the car during the same period. Both car and rail passenger growth rates were significantly below those of the airline sector in terms of domestic and intra-EU flights throughout the period.

The dominance of road transport is illustrated by a few key points. Private car ownership in the EU-15 was 232 cars per thousand inhabitants in 1975. In the EU-25 in 2003 car ownership had reached 469 cars per thousand inhabitants and continues to grow.[1] Moreover,

Table 3.1 Growth rates of main transport modes – EU-15

Freight[1]						
	Road	Rail	Inland waterways	Pipeline	Sea	Total
1970–80	4.0	0.3	0.3	2.9	5.2	3.5
1980–90	3.1	–1.3	0.1	–1.9	1.7	1.6
1990–00	3.1	–0.2	1.8	2.0	3.2	2.7
1995–04[2]	3.4	0.6	1.0	1.8	3.0	2.8
Passenger[3]						
	Cars	Bus & coach	Rail	Trams & metro	Air	Total
1970–80	3.7	2.6	1.2	0.4	8.4	3.4
1980–90	3.4	0.6	0.8	1.6	7.8	3.0
1990–00	1.9	0.9	1.3	1.2	6.1	1.9
1995–03[2]	1.9	0.5	0.9	1.5	4.9	1.9

[1] Freight figures based on tonne kilometres
[2] Figures based on EU-25
[3] Passenger figures based on passenger kilometres

Source: CEC 2004h, 2005c.

the distance travelled by all road vehicles has tripled over the same period. The strong growth in road transport has been the case in both the passenger and freight sectors as Graphs 3.1 and 3.2 show. Road freight constituted 44 per cent of all freight transport in the EU-25 in 2004 and 73 per cent of all inland modes of transport (that is, excluding shipping) (CEC, 2005c). The share of rail freight has continued its gradual decline: in 1970 rail freight accounted for 20 per cent of all freight within the EU-15. By 2004, its share had halved to 10 per cent in the EU-25 – despite the 2004 accession which brought several member states into the Union where rail played a proportionately greater role than in the EU-15. Road passenger transport includes cars, powered two wheelers and coaches: together they accounted for 85 per cent of passenger transport in the EU-25 in 2003, of which the overwhelming majority consisted of passenger cars (CEC, 2005c).

This growing dominance of road transport, whose attraction arises from its door-to-door flexibility, has resulted in a transport crisis in terms of both efficiency and the environment. Efficiency problems arise from bottlenecks, growing congestion and delays. The 2001 Transport White Paper (CEC, 2001) estimates that 10 per cent of Europe's road network is affected by daily traffic jams, increasing fuel consumption by about 6 per cent. Moreover, the White paper quotes a study that the external costs of road congestion are equivalent to

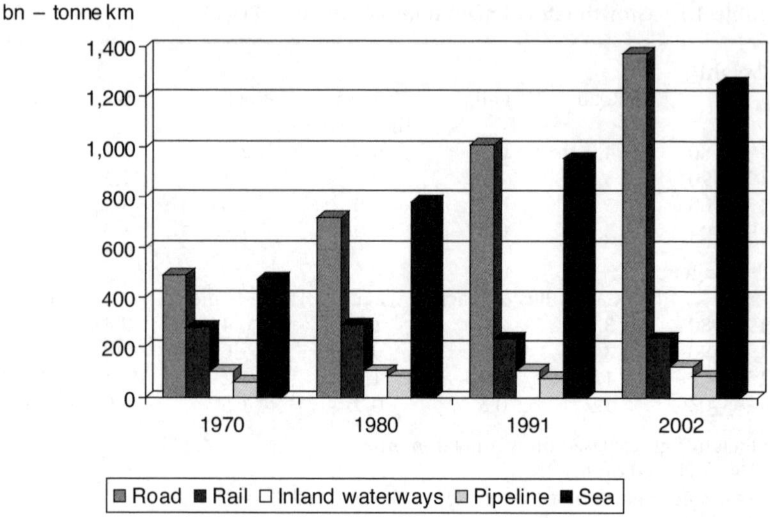

Graph 3.1 Modal performance for freight transport – EU-15 – bn tonne km
Source: CEC, 2004h.

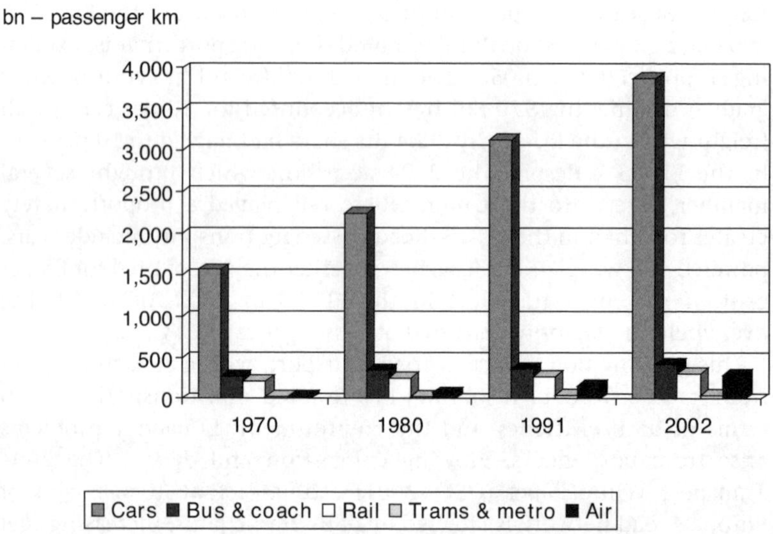

Graph 3.2 Modal performance for passenger transport – EU-15 – bn passenger km
Source: CEC, 2004h.

0.5 per cent of GDP and that, without measures to tackle congestion, its costs will rise to 1 per cent of GDP. Road transport is also one of the most environmentally damaging transport modes in relation to emissions and land usage. As a result, a major thrust of European transport policy (and the closely linked TENs policy) is to bring about a modal shift from road towards rail transport, an objective which will be immensely difficult to achieve. Policy makers will have to overcome a wide range of vested interests, change basic attitudes towards mobility and establish an environment in which investment in the rail sector is forthcoming.

Graph 3.3 demonstrates that the freight modal split varies significantly among member states, a factor which feeds into European plans to create a modal shift. In particular, it is striking how much more significant rail freight is in several of the Central and Eastern European accession states, particularly, but not only, in the Baltic states. Moreover, inland waterways make an overall minor contribution to inland freight transport within the EU (7 per cent of the total) but its contribution, for geographical reasons, is much more significant in northern Europe. The Netherlands utilises inland waterways for over 40 per cent of its inland freight needs and in Germany, for example,

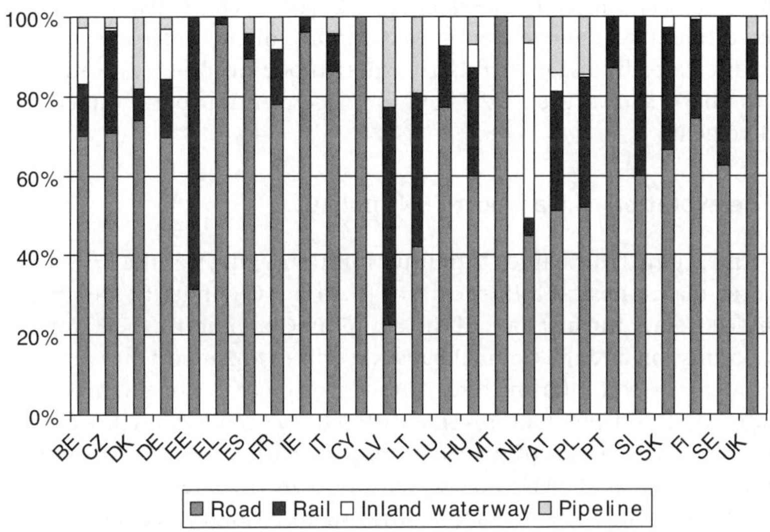

Graph 3.3 Freight modal split by member state – 2002 (tonne-km)
Source: CEC, 2004h.

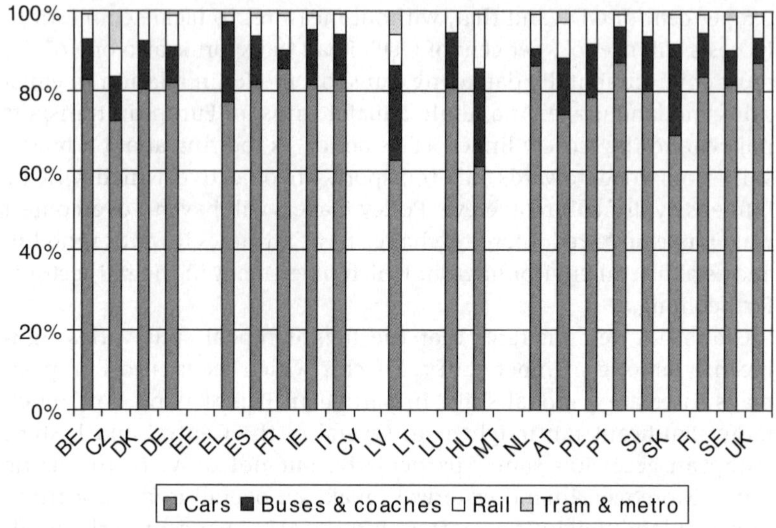

Graph 3.4 Passenger modal split by member state – 2002 (passenger-km)
Source: CEC, 2004h.

inland waterways account for about 12 per cent of freight, almost as much as that carried by the rail network. There is potential, with some investment to remove bottlenecks in the inland waterway system, to increase the usage of this transport mode heading eastwards. As Graph 3.4 shows, the modal split for passenger transport is much less divergent.

The evolution of transport TENs policy

As mentioned in Chapter 2, transport infrastructure was the first of the three main physical infrastructures to attract Community attention.[2] Indeed, since the first days of the European Community, the European Commission and the Council have recognised the importance of cross-border transport infrastructure. However, this recognition did not translate into EC-inspired improvements to European infrastructure during the first two decades of the Community's existence. The first seeds of modern transport TENs policy were sown in the late 1970s with the establishment of the Transport Infrastructure Committee (OJL, 1978) which was given wide terms of reference to discuss all matters pertinent to the development of the EC's transport infrastructure.

In November 1979, the Commission adopted a Memorandum on *The role of the Community in the development of transport infrastructure* (CEC, 1979) which aimed to promote ideas and discussion about possible Community action in the realm of infrastructure development. The Memorandum emphasised the need to incorporate infrastructure development firmly into the common transport policy (CTP), noting that 'that the common transport policy will not achieve the objectives defined for it in the Treaty and play its part in the economy as a whole unless it relates more and more to transport infrastructure' (CEC, 1979: 5).

Up to this date, any transport initiatives had been concerned with harmonisation but success had been minimal.[3] Indeed, progress had been so slow on the development of a Common Transport Policy, that in 1982, the European Parliament took the Council of Ministers to the European Court of Justice for failure to meet its Treaty obligations to develop the CTP. The Court decided in favour of Parliament in 1985. The following years were ones of relative activism in transport policy at European level, although this activism probably had as much to do with the construction of the SEM as it did with the outcome of the court case.

During the 1980s, transport infrastructure issues gradually became more pressing. International traffic was growing faster than national traffic, creating major bottlenecks. Networks were growing increasingly interdependent, undermining the conventional pattern of planning on a completely national basis. The 1979 Memorandum (CEC, 1979) also contained the seeds of the 1990s TENs programme in terms of moving towards short- and long-term infrastructure objectives (a forerunner of the 1994 transport guidelines) and of funding transport infrastructure at the Community level. In anticipation of the later subsidiarity debate, the Memorandum stated:

> Community action can in no way supplant action by the Member States, which will continue to take first place in the maintenance and development of transport infrastructure. The aim of Community action will be to co-ordinate and guide national action in such a way that the national networks as a whole can meet the future needs of Community traffic as they arise.

A key feature of the TENs programme was the need to reflect the complementarity and increased interdependence of all transport modes in the TENs programme. Thus despite their initial exclusion, by

1994, airports and seaports (although individual ports were not singled out as being of Community interest) were firmly on the TENs agenda.

At this pre-Maastricht Treaty stage, the Community lacked the legal base and the comprehensive policy framework to devise an integrated and developed transport TENs programme. However, by highlighting the need to complete the programme of transport liberalisation, the Action Programme established a firm link between access to the networks and the development of a trans-European infrastructure – a link which was made in general terms in the 1979 memorandum. The Action Programme also identified individual projects which should figure prominently in any future TENs transport network. Some of these, such as the PBKAL high speed train network and the Brenner axis, subsequently became Christophersen priority projects (see below). Standardisation initiatives, a key feature of the drive towards interoperability; traffic management, notably for air and road transport, and telematics, issues which remain central to the TENs initiative, also figured prominently in the Action Programme.

The Treaty on European Union prompted the adoption by the Commission in April 1994 of a proposal for trans-European transport guidelines. This proposal was amended in early 1995 to take account of the accession of Austria, Finland and Sweden. These guidelines outline an integrated, multi-modal transport network spanning Europe by 2010. This highly ambitious vision pulls together individual modal networks into one European transport network, a process which the guidelines envisage would stimulate 'the use of different transport modes according to their comparative advantages' (CEC, 1994: 4). This integration and interconnection of modes would, according to the Commission, take advantage of synergies and offer passengers and freight operators greater choice. This would stimulate efficiency, improve services, enhance safety and provide an optimal network from an environmental point of view.

The role of the guidelines was to fulfil the transport infrastructure requirements of the Maastricht Treaty and to facilitate the process that would lead to the integrated transport network described above. Emphasis on a 'process' implies that the mid-1990s plans for a European infrastructure would be modified and updated to take account of changing circumstances. Indeed, as seen below, the guidelines were subsequently updated and the list of priority projects have also been revisited and updated. The guidelines also required Annual Reports on their implementation and a fuller evaluation 'to verify the progress made in bringing about the progressive introduction of the network'

[Article 28 of the guidelines] within five years of the entry into force of the Decision.

The guidelines were intended to contribute to CTP objectives by helping to complete the single transport market; by contributing to the goal of sustainable mobility for people and goods throughout Europe; and by improving accessibility and strengthening social and economic cohesion. Transport TENs were to incorporate all modes and cover the whole of the territory of the Community, thereby improving the access of peripheral, island and landlocked countries to central regions. Provision was also to be made for extension of the networks to EFTA, Central and Eastern European states and Mediterranean countries.

In order to achieve the above objectives, the guidelines comprised the following elements:

- Network schemes for road, rail, inland waterways, ports, airports and combined transport and traffic management systems relating to sea and air;
- An inventory of broad measures needed for the construction of a trans-European transport network. These included the development of network structure plans, with maps; the identification of projects of common interest; interoperability; financial aid; research and development; the promotion of co-operation among interested parties; the conclusion of agreements with third countries; and any other measures needed to achieve the overall objectives of the guidelines;
- Criteria and procedure for the identification of projects of common interest.

The work of the intergovernmental Christophersen Group, which attracted more attention than the guidelines, fell outside this framework.[4] This Group identified 14 priority projects (see Table 3.2), all contained in the proposed guidelines, which it presented to the December 1994 Essen Summit and which were confirmed by the Summit as priorities. A further 21 projects 'of further importance' were also identified.[5] The Christophersen Group deliberately limited the number of priority projects to provide focus and to boost the enhanced political impetus behind the projects. Priority projects, therefore, had to be of common interest, large-scale, economically viable and provide scope for private involvement, support European Union objectives such as social and economic cohesion, respect other EU policies such as environmental protection and be at an advanced, i.e. mature, stage

Table 3.2 Christophersen priority projects (also known as 'the Essen projects')

	Provisional costings ECU million
1. High-speed train/combined transport North-South	
Berlin-Halle/Leipzig-Erfurt-Nuremberg	8,430
Brenner axis: Munich Verona	12,400
2. High-speed train: Paris-Brussels-Cologne-Amsterdam-London	
Belgian section	3,734
Dutch section	2,740
German section	3,950
UK section	5,239
3. High-speed train South	
Madrid-Barcelona-Perpignan-Montpellier	8,430
Madrid-Vitoria-Dax	4,500
4. High-speed train East	
Paris-eastern France-southern Germany	4,460
5. Conventional rail/combined transport	
Betuwe line	3,291
6. High-speed train/combined transport	
Lyons-Turin-Milan-Venice-Trieste	13,550
7. Greek motorways: north-south; east-west	5,860
8. Motorway: Lisbon-Valladolid	1,070
9. Conventional rail: Cork-Dublin-Belfast-Larne-Stranraer	238
10. Malpensa airport, Milan	1,047
11. Fixed rail/road link: Sweden-Denmark (Oresund)	6,431
12. Rail/road: Nordic triangle	4,400
13. Road: Ireland-UK-Benelux	2,920
14. Rail: UK – West Coast main line	600
TOTAL	93,290
Projects of 'further importance'	
1. Combined transport projects	3,000
2. Spata airport, Athens	2,000
3. Berlin airport	4,100
4. Maurienne motorway	1,000
5. Maratea-Elvas motorway	396
6. Danish high-speed train	1,800
7. Transappenine highway: Bologna-Florence	3,158
8. High-speed train/combined transport: Danube axis	4,700
9. Motorway: Nice-Cuneo	1,900
10. Fehmarn Belt fixed link	4,500
11. Motorway: Bari-Brindisi-Otranto	290
12. Inland waterway: Rhine-Rhône link	2,500
13. Inland waterway: Seine-Scheldt	1,350
14. Inland waterway: Elbe-Oder	650
15. Danube upgrading	700

Table 3.2 Christophersen priority projects (also known as 'the Essen projects') – *continued*

		Provisional costings ECU million
16. High-speed train: Randstad (Dutch)-German border		1,560
17. Roads: Valencia-Zaragoza-Somport		1,229
18. High-speed train: (Brenner)-Milan-Rome-Naples		8,300
19. Transrapid magnetic levitation train		4,700
20. High-speed train; Luxembourg-Brussels		85
21. Road: Naples-Reggio di Calabria		1,450–3,050
	TOTAL	49,368–50,968

of development. The list itself was not definitive and the Group wished it to be seen as part of a 'rolling plan' according to which the status of the networks remained under review with amendment of the list occurring as appropriate.

The 1995 financial regulation, the 1996 guidelines and the Christophersen Report provided the basic policy platform for furthering transport TENs. A Report on the implementation of the guidelines for the period 1998–2001 (CEC, 2004a) reported on 'solid progress' whilst urging that much more needed to be done. One positive feature of this period was considered to be the increasing share of TENs investment going to the rail sector compared to road. Nevertheless, total overall investment was seriously inadequate for the needs of a trans-European transport network.

Accordingly, a renewed interest in policy formulation began to take shape. In 2001, the transport TENs guidelines were extended to cover seaports, inland ports and intermodal terminals (OJL, 2001d). Also in 2001, the Commission adopted a White Paper on Transport Policy (CEC, 2001) which set out the main problems facing Europe's transport system and placed transport TENs within this broader context, stressing the damage being done to Europe's competitiveness by congestion, bottlenecks and missing links. In particular, it highlighted the fact that only one-fifth of the infrastructure projects in the 1996 guidelines had been carried out and that, although national sections of other projects had been completed, the trans-Europeanisation of these projects remained elusive. Moreover, the funding available from Community sources was regarded as grossly inadequate. In response to the transport challenges facing Europe, the White Paper established a number of guidelines, the following of which are infrastructure related:

- Building Transport TENs: the White Paper set out the necessity for completing trans-European projects that had already been identified. As such, it identified the Commission's intention to revise the TENs guidelines (see below). In particular, the addition of new priority projects; the raising of the maximum Community contribution to 20 per cent of the total cost of some projects; proposals aimed at developing 'motorways of the sea' and proposals linking new member states better to the EU-15 were proposed;
- Revival of the railways: emphasis on policy is to be on shifting the modal balance more towards the less environmentally damaging rail sector. The Commission proposes to do this by completing the opening of rail markets and by further measures to improve interoperability and safety. Discussion has also begin on the development of dedicated freight networks;
- Promotion of sea and inland waterway transport: the development of these two modes is seen as a way of relieving congestion on the Community's roads and of compensating for limited rail infrastructure;
- Intermodality: improved intermodality is regarded as a further way of putting forward alternatives to road transport. Policy priorities will therefore focus on technical harmonisation and interoperability between systems, especially for containers;
- Adoption of effective transport charging: the failure to incorporate full internal and external costs into the cost of transport is regarded as a prime mover in the distortion of competition between modes and the generation of pollution within the Union.

Closely following the White Paper, a High Level Group (HLG) chaired by former Transport Commissioner Karel Van Miert and comprising one representative from each member state; an observer from each of the 12 states due to accede in 2004 and 2007 and a representative from the European Investment Bank was mandated to identify by the summer of 2003 the priority projects for the development of Transport TENs to 2020. The HLG, which paralleled the earlier work undertaken by the group under Hennig Christophersen, also took into account the measures needed to support the construction of TENs and fed into the work on the subsequent review of the

Transport TENs guidelines. The HLG confirmed the need for the review because of increasing congestion (resulting from insufficient interoperability and persistent and growing bottlenecks) and missing links and the need to incorporate states from the forthcoming enlargement into European transport networks. In particular, the HLG noted the need to support the extension of the single market and greater economic and social cohesion.

The HLG had two major tasks:

1. To select a limited number of priority projects;
2. To study the financial, legal and administrative obstacles to the construction of these projects.

In the HLG's view, the identification of the label 'priority project' should ensure the concentration and co-ordination of financial resources from the Community and national sources on them and serve as an indicator for the lending policy of the European Investment Bank.

Much of the thinking in the Van Miert Report (High Level Group, 2003) was also reflected in the revised transport TENs guidelines which came into force in 2004. The HLG noted that of the 14 Essen priority projects, only three[6] had been finished and that only five would be completed by 2010. Of the remaining Essen priority projects, significant progress was expected to be made on key sections of them by 2010. The outstanding sections were to be integrated into new priority projects with a time horizon of 2020. After considering the Essen projects and proposals for 100 new priority projects, the HLG categorised the projects into four lists:

- *List 0: Priority projects in the process of completion:* these are the Essen priority projects and, although several sections of them will be completed before 2007, the HLG recommended that Community budget funding should be allocated to them in the next budgetary perspective;
- *List 1: Priority projects to start before 2010:* these projects are clearly defined with a high European value-added and the member states involved have given firm commitments to begin work on all sections of the projects by 2010. In the HLG's view these projects were to be considered the priority projects for 2007–20 and should be identified as such in the forthcoming review of the Transport TENs guidelines;

- *List 2: Longer-term priority projects:* this list includes other high European value-added projects but there was no firm commitment from the relevant member states that construction would start before 2010;
- *List 3: Other important projects for territorial cohesion:* this list was compiled to reflect projects that would contribute to economic and social cohesion, particularly, but not exclusively, on improving the access of the new member states to major European transport axes.

In recognition of the constantly evolving needs of Europe's transport networks, the HLG recommended that the list of priority projects should be reviewed more often than the ten years that had elapsed since the drawing up of the Essen list. Accordingly, the HLG recommended that a similar group be established by 2010 at the latest: this would give time to prepare proposals for new guidelines prior to the following budgetary perspective.

In relation to the HLG's second major task (the study of financial, legal and administrative obstacles to TENs), the HLG stated that greater financial resources needed to be made available from Community resources and recommended operational coordination between states for projects on a single axis, thereby foreshadowing proposals made by the Commission in 2005, and suggested that joint procedures for transnational enquiries be developed.

The HLG Report, which was published in June 2003, was the prime input into the Commission's deliberations regarding the drawing up of a new set of Transport TENs Guidelines which were officially adopted in April 2004 (OJL, 2004a). The final definitive list of priority projects (see Table 3.3) was only marginally different from that proposed by the HLG. The representatives from Belgium and Luxembourg had not approved the HLG's priority list because the upgrading of rail links between Brussels and Luxembourg did not appear on List One. Similarly, Greece had wanted the Ionian-Adriatic intermodal corridor included in List One rather than List Three. By the time the Guidelines were approved, these two projects had become priority projects and all member states were satisfied with the outcome.

In addition to financial proposals, the Guidelines also proposed the appointment of European co-ordinators to act on behalf of the Commission to encourage co-operation among all parties involved in projects along a particular axis, including users, operators, investors, etc., so that projects can be better synchronised. Consequently, in July 2005, the

Table 3.3 Priority TEN transport axes and projects

Project	Sector	Total estimated cost (at end 2004) € bn	Comments
1. Berlin-Verone/ Milan-Bologna-Naples-Messina-Palermo	Rail	45.6	European Co-ordinator – Karel Van Miert
2. Paris-Brussels, Cologne-Amsterdam-London (PBKAL)	Rail	17.5	Yes
3. HST axis – south west Europe		39.7	European Co-ordinator – Etienne Davignon
4. HST axis east	Rail	4.4	
5. Betuwe line	Rail	4.7	Yes
6. Lyon-Trieste-Ljubljana-Budapest	Rail	37.6	European Co-ordinator – Loyola de Palacio
7. Igoumenitsa-Athens-Sofia-Budapest	Road	15.5	
8. Portugal-Spain-rest of Europe	Multi-modal	12.8	No
9. Cork-Dublin-Belfast-Stranraer	Rail	0.4	Yes
10. Malpensa Airport	Air	1.4	Yes – completed 2001
11. Øresund fixed link	Road and rail	4.2	Yes – completed 2000
12. Nordic triangle	Road and rail	10.9	
13. UK-Ireland-Benelux	Road	4.5	
14. West coast main line	Rail	10.9	Yes
15. Galileo	Satellite navigation	3.4	
16. Sines/Algeciras-Madrid-Paris	Freight rail	6.1	

Table 3.3 Priority TEN transport axes and projects – *continued*

Project	Sector	Total estimated cost (at end 2004) € bn	Comments
17. Paris-Strasbourg-Stuttgart-Vienna-Bratislava	Rail	10.1	European Co-ordinator – Péter Balázs
18. Rhine/Meuse-Main-Danube	Inland waterway	1.9	
19. Iberian peninsula interoperability	Rail	22.3	
20. Fehmarn belt	Rail	7.1	
21. Motorways of the sea	Maritime	Individual projects to be announced by 2007	
22. Athens-Sofia-Budapest-Vienna-Prague-Nuremburg/Dresden	Rail	11.1	
23. Gdansk-Warsaw-Brno/Bratislava-Vienna	Rail	5.5	
24. Lyons/Genoa-Basle-Duisburg-Rotterdam/Antwerp	Rail	22.7	
25. Gdansk-Brno/Bratislava-Vienna	Road	7.8	
26. Ireland-UK-continental Europe	Road and rail	4.7	
27. 'Rail Baltica' Warsaw-Riga-Talinn-Helsinki	Rail	2.7	European Co-ordinator – Pavel Telicka
28. 'Eurocaprail' Brussels-Luxembourg-Strasbourg	Rail	1.4	
29. Ionian/Adriatic intermodal corridor	Rail	2.5	
30. Seine-Scheldt	Inland waterway	2.5	

Source: CEC, 2005b.

Commission published a draft decision proposing the appointment of six European co-ordinators to fulfil this role (CEC, 2005a). The co-ordinators are charged with:

- Promoting joint methods to evaluate projects, including, if necessary, advice on the financial package;
- Drawing up an annual progress report on the project and on factors that could affect or delay the project;
- Consulting all interested parties to obtain a better understanding of the demand for transport services and the likelihood of obtaining funding for such services.

The proposed co-ordinators (see Table 3.3) include three former European Commissioners – Karel Van Miert, Etienne Davignon and Loyola de Palacio. Five of the co-ordinating roles relate to specific priority projects. The sixth project is concerned with deployment of the ERTMS rail traffic management system in the railway corridors. Given the technical expertise required for this role, the Commission have designated Karel Vinck, on secondment from SNCB and a former president of the European Railways Community.

Overall, the EU's TENs initiative represents an attempt to move away from a process of infrastructure development based entirely on national priorities, a process that appears increasingly outdated given the increasing economic interdependence across Europe. However, despite the erosion of the national monopoly on infrastructure development, its extent should not be overestimated: it is clear, for example, that whilst transnational concerns are embedded in the list of priority projects, the member states themselves have exercised a strong influence over the choice of projects, including ensuring that their projects were included on the lists with the highest priority. The degree of control that states continue to exercise varies by transport mode and is explored more fully in the following section on individual modes.

Roads

Given the congestion currently prevalent on Europe's roads, roads lie at the heart of Europe's mobility gap. However, although roads are easily the most dominant form of transport in Europe, they do not dominate the Transport TENs programme. Only three of the 30 TENs-T priority projects exclusively involve roads: their primary objective is to improve road connections between the periphery and centre of the EU.

A further four projects are intermodal in nature and their main purpose is to facilitate the switch of passengers and freight from one form of transport to another (see Table 3.4).

This relatively minor role for road transport in the TENs initiative, given the dominance of this form of transport, is in line with broader

Table 3.4 Road-related TEN-T priority axes and projects

Motorway axis South-east Europe	Athens-Sofia-Budapest and Igoumenitsa-Patras Significant improvements to road network of south-east Europe in terms of quality of road and improved linkages to ports, airports, other roads and major cities. Partially completed – development of a Christophersen project.
Motorways Spain and Portugal	Continuing improvements to motorway networks of the Iberian Peninsula, involving construction of 2,265 km of new motorway. Enhances links between the centre of Europe and its peripheral regions. Part of a multi-modal package of projects.
Road-rail Denmark-Sweden	Oresund fixed rail-road link between Denmark and Sweden. Four-lane motorway and double-track railway line as link and access routes – into service in 2000.
Multi-modal Nordic triangle	Upgrading of road, rail and maritime infrastructures to improve transport between the Oresund fixed link, Stockholm, Oslo, Turk-Helsinki and the Finnish Russian border. Includes 1,900 km of roads.
Road	Ireland-UK-Benelux road link. Links Cork, Dublin and Belfast by road and ferry with Wales and Scotland and through England to Felixstowe-Harwich and hence with Benelux road network. Some new building (largely in Ireland) and some upgrading – involves 1,500 km of roads in total.
Motorway axis	Gdansk-Brno/Bratislava-Vienna – the construction of a largely new motorway (some upgrading) from the Baltic to Central Europe. Involves 1,038 km of road. Poland has one of the least developed road networks in Europe and has lost out in terms of FDI because of poor infrastructure.
Road-rail axis	Ireland-UK-continental Europe. Linked to other TEN-T axes, this axis includes Hull as an exit/arrival point to the UK, in addition to Felixstowe.

transport policy objectives which are aimed at 'revitalising alternative modes of transport to roads' (CEC, 2001) and at encouraging a modal shift away from road. It is the increasing congestion, greenhouse gas emissions and the wasteful land usage[7] of roads that has pushed this mode of transport down the list of policy objectives. Given road transport's advantages in terms of flexibility, especially its ability to transport goods and passengers from door-to-door, enticing users away from this form of transport will be far from easy.

There are two main strands to the EU's road network policy, which parallel the approach to other sectors and transport modes: the construction of new infrastructure and the upgrading of existing networks; and measures to ensure that the provision of goods and passenger services across the road infrastructure is as efficient and as of high quality as possible. Moreover, the thrust of the policy has been, through a variety of measures, to establish a competitive single European area for the transport of goods and passengers by roads, in the same way as it has been to establish single markets for other transport sectors.

The construction of a trans-European road network is fraught with problems. Road congestion is a serious and growing problem with car ownership levels continuing to rise. A trans-European transport network implies greater coherence and continuity on international routes. Roads with a purely local or regional impact are not strictly speaking part of that network. However, improved local systems can ease congestion resulting from intermingling of short and long distance traffic. In addition, construction of local roads can improve links with other transport modes and improve access to other parts of the transport network, further reducing pressure on the road network.

New building and upgrading projects within transport TENs have been selected for their contribution to major routes, to the elimination of bottlenecks and to the cohesion objective of the EU. Significantly, and unsurprisingly given their low base, it is the peripheral countries that have extended their networks the most. In 1990, for example, only 25 kilometres of motorway were operational in Ireland (CEC, 2005c): the Irish section of the Ireland-Benelux corridor involves the construction of 360 kilometres of motorway between 1996 and 2010. In 1990, Portugal possessed only 316 kilometres of motorway: by 2003, this had risen to 2,000 km, of which a significant proportion are integral to transport TENs projects (CEC, 2005c). The Greek motorway system was 190 km long in 1990, a total which had quadrupled by the end of the decade, partly because of TENs, and which will increase further as TENs projects near completion. The length of Spanish motorways has more than doubled since 1990: far from all the new capacity is transnational in nature but progress in plugging gaps in the system improves the overall integrity of the network. Among the 2004 accession states, it is Poland that has the least developed motorway network: the completion of the Gdansk-Katowice section of the Gdansk-Brno/Bratislava motorway axis will itself more than double the length of the Polish motorway network. Elsewhere in the EU, the motorway network is much denser and construction work has focused on easing congestion and removing bottlenecks.

In relation to creating a single European area for the transport of passengers and goods by road, the European Union has taken a range of measures in the following categories:

- Access to the profession: any passenger or freight transport services company based in a member state has the right to supply services to any other member state on an equal basis with nationals of the member state provided they meet the criteria set out in directives relating to repute, financial capability and professional competence;
- Freedom of movement: any operator of goods and passenger services can operate freely throughout the Union without discrimination on the grounds of nationality or place of establishment provided they are in possession of a Community licence issued by the home member state, renewable every five years. Freedom of movement has been helped by the removal of the restrictions on cabotage across all modes with the result that operators are not only able to carry passengers and goods between member states of the Union but also within the borders of individual states;
- Safety and social aspects: common standards have been established for driving times, working hours and rest periods to justify minimum safety standards and to ensure that the enhanced competition that the opening of markets entails does not lead to operators taking shortcuts on safety. These standards are enforced by the use of tachographs and other means of inspection;
- Taxes and charges for the use of infrastructure: taxes on coaches and lorries, fuel taxes and charges for the use of infrastructure through tolls, etc. contribute to a complex mixture of charges throughout the EU which lack transparency and can distort competition. A key objective of the Transport White Paper is to develop a system of infrastructure charging that truly reflects the costs of infrastructure, including all external costs such as accidents and all types of pollution, including noise, air pollution and congestion. A scheme which

fully meets all EU objectives would incorporate all transport modes and types of users. Attainment of this objective remains some way off but the Commission has made a start with its 2003 proposal for a directive on the charging of heavy goods vehicles for the use of certain infrastructures.

Overall, the combination of measures to improve European road networks, particularly in terms of improving the connectivity between national road networks, and the initiatives to remove obstacles to providing services throughout the Union has played an important part in encouraging both transport providers and users to reassess how they operate and their logistics and distribution networks. Moreover, the case of roads demonstrates that the process of developing infrastructure is not only about physical construction but also concerns agreements on common standards and liberalisation of services, especially in terms of improving access to services. In short, the value of networks extends beyond mere physical provision into measures to improve the efficiency of its usage. This implies that the policy/corporate strategy interface has been shaped, in part, by the objective of improving mobility for business. One by-product of the changes has been the emergence of private businesses who are transforming themselves into infrastructure business through the ownership and control of roads and nodes such as airports and seaports. Such a development is only possible because of the relaxing of aspects of the control model. These issues are explored more fully in the final section of this chapter.

Rail

Rail TENs lie at the heart of the transport TENs initiative and have always done so. Rail projects, for example, comprised 69 per cent of the investment in the projects outlined in the original TENs transport guidelines (CEC, 1995) and eight of the 14 Essen projects with a further two involving road-rail links. Out of the 30 priority projects emerging from the work of the High Level Group under Karel Van Miert, 18 are entirely focused on developing trans-European rail links; a further three have a road and rail component and another is a multimodal project with a rail component (CEC, 2005b). The ultimate intention is that by 2020, the transnational rail network will amount to 94,000 km of track, of which around 20,000 km will comprise track suitable for speeds of at least 200 km/h (CEC, 2005b). Moreover, attention also needs to be paid to ensuring the interconnectedness of more conventional track to enhance the attractiveness of the rail option, particularly *via* the promotion of intermodalism.

The attraction of rail in addressing the emergent mobility gap is that utilisation of the full potential of rail networks would ease congestion on Europe's roads and reduce carbon dioxide emissions which are much lower per passenger kilometre for rail than for other modes of transport. Rail also uses land more efficiently than road-based transport. It is essentially for these reasons that the current emphasis, not only of the TENs project but also of the Common Transport Policy is to engineer, if possible, a shift in the modal balance away from road transport towards rail in particular, but also to inland waterways and short sea shipping where feasible.

A genuine transnational rail network would enable high speed trains to transport freight and passengers throughout Europe using the same equipment and personnel. In order to achieve this, two key objectives need to be pursued relentlessly – market access and interoperability. However, the rail sector has not loomed large in European transport policy for long. Although air transport, road haulage and the maritime industry were included in the 1985 Single Market White Paper (CEC, 1985), rail was omitted from this seminal document. This neglect continued until the late 1980s when the Community for European Railways, the association for rail operators in the EU, put together plans for a European network (CER, 1989) and the European Commission brought forward its first comprehensive rail policy statement. The resulting directive (OJL, 1991) came into force on 1 January 1993. This first attempt to open the rail market seemed significant at the time but in retrospect, it did little to change the status quo. By requiring the separation of accounts for infrastructure and *via* provisions regarding autonomy, it did at least set the foundations for later initiatives to open international markets.

The next attempt to open Europe's rail markets occurred in 1995 but went nowhere because of lack of support in the Council. Although it has changed somewhat in some member states, Europe's rail sector has long been characterised by heavily indebted and subsidised state monopolies. In 2000, for example, the Commission estimated that state subsidies for rail infrastructure approached €25 billion and that compensation for public service obligations was almost €10 billion (Vinois, 2002). This plus the vested interests of incumbents and the slowness of liberalisation helps explain the falling share of rail in traffic throughout Europe.

An unanticipated consequence of the Treaty of Amsterdam was facilitation of the opening of Europe's railways. By making transport subject to co-decision, the Amsterdam Treaty conferred a greater role in the formation of transport policy to the European Parliament, an institution which was anxious to push European transport policy further. In 1998, the Commission tabled three proposals on rail infrastructure (commonly referred to either as the 'first rail package' or the 'railway infrastructure package') and, following a long and controversial debate during which the package was considerably strengthened as a result of European Parliament interventions, the three directives eventually came into force on 15 March 2003 (OJL 2001a, b and c). In essence, this package opened international freight to competition on a significant part of the Union's network. Market opening had to overcome the power of the incumbent operators and the directives sought to do this by separating out 'essential functions' to ensure transparent and non-discriminatory access to the infrastructure, including most importantly, separation of control over the track and the services that operate over the track. In order to monitor developments and to provide an efficient appeal system, member states were also required to set up an independent regulatory body to rule on complaints.

In 2002, the Commission brought forward its second rail package with a view to take the construction of an integrated rail area further. This package contained five proposals designed to improve safety, interoperability and the extension of market opening to domestic rail freight, including the introduction of cabotage. Moreover, the package proposed the establishment of a European Rail Agency, subsequently set up in Valenciennes in France, to direct work on safety and interoperability. This second package came into force in April 2004 (OJL, 2004b, c, d, e) causing the then Transport Commissioner Loyola de Palacio to declare:

> This is the end of the physical and technical barriers in European freight railway transport: this new context will change radically the picture for rail transport and will really boost it. It will also contribute to the fight against road congestion. This is a revolution which represents a genuine European rail transport integration (Commission Press Release, 22 April 2004).

In 2004, the Commission took things a stage further when it tabled proposals for a third rail package, comprising a Communication, four legislative measures and a Working Document on an extended impact assessment for the gradual opening of the market for international passenger services (European Commission, 2004b, c, d, e, f, g). The main thrust of the third package measures is to open the market for rail passenger services by 2010; to protect the rights of international passengers; to introduce minimum quality clauses in rail freight contracts and measures to attest to the qualifications and standards of locomotive drivers. Significant progress on the third package was made in December 2005 when the Transport Council reached political agreement on the proposals.

As important as market access is in securing a genuine transnational rail network, it is the issue of interoperability which is intended to facilitate seamless transition from one network to another. Attainment of the goal of interoperability will help achieve a number of CTP objectives, including greater mobility and deeper economic integration; more effective and efficient use of rail infrastructure which will contribute to the goals of modal shift and of limiting the environmental impact of mobility; enhanced safety through the use of one trans-European rail control, signalling and telecommunication system rather than the multitude of systems in current use; and create a genuine single market in rail equipment.

The concept of interoperability was introduced in the Maastricht Treaty. Directive 96/48/EC spells out the concept more specifically in relation to the high-speed train network and refers to it as 'the ability of the Trans-European high-speed rail system to allow the safe and uninterrupted movement of high speed trains which accomplish the specified levels of performance. This ability rests on all the regulatory, technical and operational conditions which must be met in order to satisfy essential requirements'. Technical fragmentation and the challenges of interoperability have posed particularly large obstacles in the rail sector, including significant differences in gauges, electric current, maximum axle loads for wagons and locomotives, traffic management system, requirements for staff, etc.

Solutions to these problems are gradually evolving. In Directives 96/48/EC and 2001/16/EC dealing with high-speed and conventional rail respectively, the EU has laid down the essential requirements for interoperability at network interfaces. The Directives provide the framework for technical specifications for interoperability (TSIs). The first TSIs (OJL, 2002)[8] were drawn up on the basis of recommendations of experts working for the European Association of Rail Interoperability (AEIF) which was set up by the UIC, UNIFE and UITP precisely for this purpose and came into force in December 2002. This first group of TSIs sets out common specifications for future infrastructure, rolling stock, energy supply, signalling, maintenance and operations. Following the coming into force of the second rail package and the subsequent establishment of the European Railway Agency, the responsibility for developing TSIs has passed from the AEIF to the new body.

Work on conventional rail interoperability is continuing. At the time of writing, TSIs relating to telematic applications for freight (OJL, 2006a) and to noise from rolling stock had been adopted (OJL, 2006b). Others concerning freight, control command and signalling, traffic operation and management, rail tunnel safety and access for people with reduced mobility are in preparation. The first batch of TSIs for both high-speed and conventional rail are only the beginning. Revised and new TSIs are to be anticipated and the European Rail Agency intends to expand its activities into systems for certification of rolling stock maintenance workshops and for identification and registration of rolling stock. It also proposes to bring forward recommendations on vocational competences, particularly for drivers and trainers.

Overall, significant progress has been made since the mid-1990s in providing the framework at least for the development of trans-European rail networks. Market access has been established, although it remains to be seen how enterprises will ultimately respond to these opportunities and the interoperability initiative has moved on from developing general policy to drawing up detailed requirements for interoperability. Market access and interoperability in themselves will not create a trans-European rail network but without them, rail TENs would definitely not materialise. The realisation of the vision of a seamless, trans-European rail network requires a number of other factors to become a reality. These include, for example, resolution of infrastructure pricing issues and new financial initiatives. There is a view that creation of an exclusively trans-European freight network would be beneficial but there are, as yet, few signs of this happening.

Inland waterways

Inland waterways, including rivers and canals, proved influential in the early industrial development of Europe. Their role declined following the emergence of quicker transport modes, first rail and later roads, and the subsequent lack of maintenance. The use of larger vessels also made certain secondary routes unattractive. However, there has been a resurgence of interest in inland waterways following the launch of both the single market and the TENs programme. Excess capacity, low energy consumption and the potential for inland ports to operate as intermodal nodes, in part, explains the renewed policy emphasis on inland waterways which has cumulated in 'NAIADES'[9] (CEC, 2006a) and have made inland waterways seem an attractive option in the push for an integrated European transport system. In particular, the potential for connecting inland waterways to other modes of transport provides scope for inland waterways to fill some of the mobility gap in northern Europe.

At present, Europe has over 30,000 km of canals and rivers and a fleet of 12,500 vessels, roughly equal in capacity to 10,000 trains or 440,000 trucks. Traditionally, inland waterways have carried bulk cargoes, such as coal or iron ore, over long distances. In the 1990s, new markets began to emerge involving container traffic, intermodal transport and the carriage of specialised cargoes such as hazardous chemicals and refrigerated goods.

In reality, the development of an integrated inland waterway system will involve a limited number of member states. The core network currently links the Netherlands, Belgium, Luxembourg, France, Germany and Austria *via* a series of navigable main rivers with secondary waterways extending the reach of the main network to hundreds more locations. The 2004 enlargement, however, opens up the possibility of the Danube becoming a key east-west waterway link between the Black and the North Sea.

The concentration of the network means that, although the share of inland waterways in total EU freight transport is relatively low at 7 per cent (CEC, 2004h), its share in the core network is much higher (see Graph 3.5) in those countries that utilise inland waterways for the

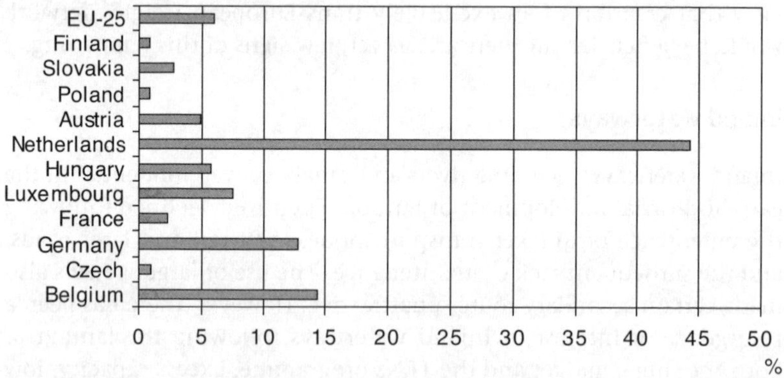

Graph 3.5 Inland waterways as % of freight by country
Source: European Commission, 2006.

carriage of goods. In the case of the Netherlands, for example, where the large sea ports are well served by an extensive hinterland of inland waterways, inland waterways account for almost 45 per cent of freight transport. In other regions, such as northern France, the role of inland waterways in freight transport is greater than the national figures alone suggest.

Inland waterways were included in the early TENs proposals and the 1994 Transport Guidelines (CEC, 1994) drew on the work of the Transport Infrastructure Committee and on the Commission's 1992 proposals for a European inland waterway network (CEC, 1992). At the heart of the proposed network were four main axes, axes which reflected and increased the interconnection of the already existing network of northern Europe. These were:

1. The Rhine axis connecting Dutch and Belgian ports to the industrial heartlands of western and southern Germany, eastern France and Switzerland;
2. The north-south axis formed by the rivers Meuse and Scheldt and including the rivers Rhone and Saone in the south, thereby improving links between the seaports of Belgium, Netherlands and Dunkirk with the rest of Belgium, the Paris basin and the Moselle and Rhone;
3. The east-west axis through northern Germany, starting at the seaports of Hamburg and Bremen and connecting with Germany's inland waterway network. This axis links Belgian and Dutch ports in the west to the waterways of the east by the Elbe and the Mittellandkanal;
4. The south-east axis formed by the Danube which, *via* the Main-Danube canal, provides a link with the Rhine.

Despite, identification of the configuration of the main features of the network, inland waterways were not included in the Christophersen priority projects, although they did figure in supplementary Christophersen projects. However, two inland waterway projects were included in the priority projects in the 2004 TENs guidelines (OJL, 2004a). These are the:

- Improvement of the link between the Rivers Seine and Scheldt to connect the Paris region and the Seine basin within the inland waterway network of Benelux;
- Removal of bottlenecks on the Rhine/Meuse – Maine – Danube corridor to improve navigability and ultimately to open up inland waterway traffic between the North and the Black Seas.

Like all other networks, creation of the physical side of the network is only part of the process of shaping a seamless, interconnected network. Other obstacles remain. In relation to inland waterways, as with all other TENs-related sectors, liberalisation is a pre-condition for interconnectedness. Indeed, all the key liberalisation measures for inland waterways were in place by 2000: obstacles to market access had been removed, for example; prices were determined by the market and key harmonisation measures were in place such as harmonisation of the conditions for obtaining a boatmaster's certificate and for the technical requirement of vessels. However, the following obstacles to the creation of a genuine pan-European inland waterway network remain and have influenced the drawing up of NAIADES.

First, the inland waterway sector is characterised by a large number of private sector ship owners and competition is strong. Although increasing competition has been the watchword of European transport policy, the market structure has rendered reinvestment difficult. Secondly, inland waterway vessels have a long life, limiting the opportunities for modernisation. Thirdly, the image of inland waterway is of an old-fashioned, slow moving mode of transport. This creates problems, both in terms of attracting new business to the sector and in attracting labour: career prospects are seen as limited and there are problems of succession in family-owned businesses. Fourthly, the network is fragmented by virtue of the patchwork of institutions which have jurisdiction over the network. These include, *inter alia*, the EU, the Central Commission for Navigation on the Rhine, the Danube Commission, the United Nations Economic Commission for Europe and the European Conference for Ministers of Transport. Finally, and most directly connected with physical infrastructure, is the finance needed to eliminate strategic bottlenecks within the network and to maintain the waterways themselves.

The NAIADES programme provides some responses to the above constraints on developing an integrated European inland waterways network. These are framed around the following five strategic areas to be developed between 2006–13:

1. The creation of favourable conditions for service and the attraction of new markets: the positioning of inland waterways as key nodes in intermodal transport requires the nurturing of synergies between freight forwarders, shippers and ports and the development of a critical mass. The NAIADES Action Plan intends to do this by introducing measures to help overcome start-up barriers,

to harness entrepreneurship by facilitating access to finance and by improving the administrative and legislative framework to establish and guarantee a level playing field among member states, within the inland waterway network and between modes of transport;
2. Measures to foster fleet modernisation and innovation: the aim is to improve the efficiency of logistics and the environmental and safety performance of inland waterway transport;
3. Measures to resolve the problem of skills shortages. Greater mutual recognition of qualifications and increased investment in human capital, particularly in devising a common framework for education and training standards are regarded as central to resolving the skills problem in this sector;
4. The promotion of inland waterways. The Action Plan claims that the potential of inland waterways to play a central role in European logistics planning is not fully understood and an important of activities supported by the Plan will be measures aimed at promoting an updated and positive image for inland waterways;
5. The provision of adequate physical infrastructure. This strategic area lies at the heart of the trans-European network area and is intended to help the removal of bottlenecks that are preventing inland waterways from realising their true potential. Such bottlenecks include restrictions on draught, bridge clearance and the dimension of locks. More specifically, the Action Plan incorporates support for the above two TENs priority projects and other projects of common interest, including port and terminal development; plus the appointment of a European co-ordinator to help oversee these projects; greater priority in spatial planning to the (re)development of industrial areas near waterways to facilitate multi-modalism and greater contacts between sectors at a planning and project level. In particular, the Action Plan looks to the development of infrastructure charging to cover all transport modes. Given the safety and environmental advantage of inland waterways, this can only help their promotion. The Plan also seeks the implementation of River Information Services across the network in the planning and management of traffic on Europe's inland waterways network.

In order to support the realisation of the above strategies, the Action Plan proposes to explore options to end the fragmentation of the organisation structure governing Europe's waterways. Modernisation of this situation should improve efficiency by ending the co-existence of competing sets of rules on a range of issues.

The maritime sector

In 2004, maritime transport accounted for over 70 per cent of extra EU-25 trade by volume and nearly 20 per cent of intra EU-25 trade (CEC, 2005c). Shipping is also the most energy efficient and least environmentally hostile of the major transport modes yet, despite its potential for shifting traffic from congested land corridors, it has had a relatively low profile in terms of overall transport policy generally and of the TENs initiative itself. Indeed none of the Christophersen priority projects nor the supplementary list of projects related to shipping or to ports – the infrastructure element of the shipping industry.

One reason for the apparent neglect of the maritime sector in the TENs initiative is the large degree of competition that has long existed between ports. Ports have evolved under a variety of different traditions within the European Union[10] but competition still acts as an important factor in ensuring price keenness and efficiency and operates more extensively in this mode than in others. Therefore, EU intervention to promote individual ports as key points on European transport networks had the potential to distort this competition and undermine the basic principles of SEM and TENs policy. Policy in relation to the other component of the maritime sector – short sea shipping – was more active during the 1990s. At the onset of the SEM campaign, short sea shipping, like other transport sectors, was heavily protected, especially in southern Europe. However, as in other transport sectors, most restrictions on cabotage were lifted during the 1990s.

Attitudes to the inclusion of the maritime sector in the TENs initiative have changed dramatically since the 1990s and the sector has become more high profile and central to the TENs project. Factors underpinning the greater emphasis on maritime transport include:

1. Environmental concerns: carriage of freight by sea is more fuel efficient and freer of congestion than other modes. Congestion is possible if there are long waiting times at ports but this has not hitherto been a problem in Europe;
2. A growing general recognition of the strong growth of sea transport and its underused potential: often neglected or regarded as the poor relation among modern transport modes, growth in the quantity of freight carried by sea has been similar to that carried by road since

1990 (see Table 3.1). In total, 41 per cent of freight moved within the EU goes by sea compared to 45 per cent by road;

3. The current and forecast growth of freight transport: the European Commission anticipates that inland freight growth in the EU-15 will grow by 70 per cent in volume terms between the early 2000s and 2020. In the 2004, accession states such growth is forecast to be as high as 95 per cent over the same period. Given the already congested state of the Union's roads, such freight growth will be best absorbed *via* diversion to other transport modes: rail is a long-term possibility but short sea shipping looks the most promising option for large-scale diversion of traffic from the roads;
4. Sea freight can bypass natural barriers such as mountain ranges or connect peripheral and more inaccessible areas to the European market. For example, Greece which at present has no land links with other EU member states is directly connected by sea to Italian ports. Similarly, road links between Spain and France are funnelled into bottlenecks by the geography of the Pyrennean mountain range. Transport of goods from the ports of north-western Spain to the French Atlantic coast circumvents this problem;
5. The greater emphasis on the need to improve intermodalism: road has the great advantage over competing transport modes of providing a door-to-door service. If short sea shipping is to compete with other modes, especially road, or to attract traffic diverted from other transport modes, the point at which freight is handed to or from shipping must operate as efficiently and quickly as possible. The intermodal imperative operates across all transport modes. If the seamlessness of the connections between sea-based transport and other forms of transport (that is, at the ports) is improved, this will facilitate the emergence of holistic and multi-modal logistics chains in which shipping plays a key role. Such chains would be more sustainable and ultimately more efficient than road-only transport and improve access to markets throughout Europe.

The 2001 Transport White Paper gave the first official indication of this change of direction with its introduction of the 'Motorways of the Sea' concept (CEC, 2001). More specifically, the Commission has defined motorways of the sea as 'regular, high-capacity ferry routes between key ports in the EU' (CEC, 2005b). The long-term implications of this new concept are far-reaching. In particular, initiatives to establish new, regular maritime links that are sufficiently frequent to attract large quantities of traffic will lead to a concentration of traffic flows on selected ports and regions and facilitate the growth of intermodalism. In 2006, the identification of particular motorways of the sea is at a general level. Four major maritime corridors or 'motorways' have been included in the 30 TENs priority axes. The four are:

1. The Baltic Sea Corridor: this links Baltic member states with the member states of Central and Western Europe. Many such links are already in place but they are connected into integrated transport corridors. Moreover, the Baltic motorway poses one challenge not facing the other corridors: that is, the need to secure all-year access to peripheral areas *via* the provision of ice-breaking services;
2. The Western European motorway: this links the Iberian Peninsula *via* the Atlantic arc to ports in the Irish and North Seas;
3. The south-west Europe corridor: this aims to improve connections within the western Mediterranean area, particularly Spain, France, Italy and Malta;
4. The south-east Europe corridor: this connects the Adriatic to the Ionian Sea and to the eastern Mediterranean. Ultimately, links between the south-west and south-east Europe motorways will enhance integrated transport within the EU and the architects of the projects envisage ultimately extending this network into the Black Sea.

Two types of development are envisaged to make motorways of the sea a reality. The first is the improvement of infrastructures, not only conventional port infrastructures but also infrastructure to ensure smooth transition from sea-borne to land transport (road and rail) and to inland waterways. The second is the development of facilities, such as electronic logistics management systems, safety, security, administrative and custom facilities and ice breaking and dredging where appropriate, which will improve the efficiency of ports. These developments are intended to help short sea shipping overcome some of the obstacles[11] that have been identified by the Commission as hindering the further development of the sector (CEC, 2004a) and requires the commitment of all participants in the supply chain, not only from the port and shipping side but also from land-based transport operators, logistics companies, freight consolidators and the public sector to ensure the full intermodal potential of motorways of the sea is captured.

In order for the Motorways of the Sea to become a reality, certain key and difficult decisions need to be made.[12] In particular, the maritime links that have the potential to become motorways of the sea (that is, the demand to support regular and frequent freight services) need to be identified and the measures to concentrate the traffic need to be taken. The process of port selection raises highly delicate issues that affect local, regional and national politics (in the latter case, when the choice concerns two ports on either side of a national border). The choice of one port over another is likely to divert traffic from one port to another and, over time, could change economic development patterns. The designation of ports within the motorways of the sea project represents a direct intervention into the market and alters the forces of competition within a region (the major reason why ports were not included in the original TENs programme). The identification of individual motorways therefore should be based on clear criteria and be carried out according to a clear and transparent process. The probable outcome is that ports will compete fiercely with each other to become part of the process and the final choices will be determined as the result of a long process of political bargaining. This is particularly likely to be the case when selections are made between ports situated in different countries as states are likely to lobby for their port(s) to be favoured.

In order to qualify as a link within one of the four major corridors and to benefit from the, albeit limited, Commission support, any initiative must demonstrate that it promotes cohesion and/or modal shift and that it passes normal commercial viability tests. The overall aim is to make the targeted motorways flagships of integrated short sea shipping services that will generate a critical mass for further investment within and beyond these corridors. As with other TENs initiatives, PPPs are envisaged to implement these projects. This requires the partners to work together to target key ports where the achievement of critical mass is more likely and is based on the need to integrate logistics value chains. As the motorways of the sea concept focuses on breaking down barriers between transport modes and making better use of existing transport resources, it does not necessarily require massive investment but it does need assorted stakeholders to agree on priorities and more effective marketing of short sea shipping.

Airports

Airports have an increasingly important role to play as nodes in Europe's transport networks but, despite the growth in saturated capacity at several of Europe's major airports and the continuing fragmentation of Europe's airspace, airports are low on the list of transport

TENs priorities. Indeed, only two of the 30 priority axes and projects directly incorporate airports.[13] These are:

1. The construction of a new Lisbon airport during the period 2006–15 as part of the multi-modal axis linking Portugal and Spain to the rest of Europe;
2. Malpensa Airport: this was one of the original Christophersen projects and was completed in 2001. However, the development of Malpensa into an international hub for northern Italy and southern Europe continues. New facilities planned or under consideration include a new aircraft maintenance hangar, a third module for Terminal 1, a third runway, further investment in a logistics park and improved road and rail access to the airport.

Although the EU generally acknowledges the importance of civil aviation to Europe's overall transport networks and particularly emphasises the importance of airports in the emergence of intermodalism and of multi-modal hubs to support the development of European, national and regional logistics networks, the provision of the associated infrastructure in the form of networks is not regarded as primarily a matter for Community action. One reason for the relative absence of the airports from the TENs initiative is the view that the development of airports is primarily an issue for private investors. Indeed, the European Parliament originally opposed the inclusion of Malpensa Airport on these grounds. Moreover, it can be argued, as with seaports, that identification of specific airports as airports of Community interest represent a distortion of competition.

* * *

The foregoing analysis deals with transport on a mode-by-mode basis but plugging the mobility gap is about more than individual modes or individual pieces of infrastructure, however important. Mobility in Europe is as much about ensuring that traffic from one transport mode flows seamlessly and efficiently to another as achieving interconnectedness within individual, separate modes. It is the need for connections between as well as within networks that gives transport a layer of complexity that is not apparent in the energy and telecommunications networks. Not all actors in the transport sector have as yet adopted this holistic approach to the development of transport networks (see following section) but there are signs, especially in the private sector, of recognition of the potential gains from doing so. The emergence of multi-modal infrastructure companies and of logistics companies that are active in several transport modes to maximise the benefits for their customers are examples of this.

Stakeholders and strategic responses to transport TENs

The organisation and configuration of Europe's transport system is influenced by many factors at a variety of levels. The integrity and effectiveness of Europe's transport system in turn has important ramifications for the competitiveness and organisation of European business generally. European transport has undergone significant change in terms of regulation during the last decade of the twentieth century and the demands on transport have changed, not only as a result of increased traffic but also because of challenges from the changing business environment. The response of Europe's transportation and logistics industry to these challenges is fluid and emerging as the implications of changing policy, infrastructure development and new business conditions interact with each other and take time to roll out. Moreover, the process of change continues: as the earlier section shows, the liberalisation of Europe's rail sector is taking place later than that of other sectors, for example, and is not complete in the mid-2000s. Moreover, change comes from sources other than regulation, such as technology and traffic management systems.

The first key challenge to the business environment came from the creation of the single market. The reduction and removal of barriers to trade within Europe encouraged enterprises to regard the European market as one big market rather than a succession of separate markets. Consequently, intra-European trade grew as did the need for increased capacity of transport and, in particular, for improved transport quality in terms of the eradication of bottlenecks and missing links and for the development of seamless connections between national networks. The single market idea also encouraged manufacturers to restructure their warehousing, distribution and general logistics: this often entailed the creation of a European distribution centre with or without satellite distribution centres to service sub-regions of Europe. The success of these strategies depended on a range of factors, including liberalisation of European transport policy and improvements to trans-European networks.

The second challenge emanates from the 2004 enlargement which extended the SEM eastwards and encouraged business in general to look for new low cost locations with potential new markets in the east.

In order for enlargement to fulfil its potential, major improvements in transport infrastructure are essential given the poor quality and quantity of infrastructure in several countries and the low number of east-west links.

The third challenge is globalisation which has increased trade generally and has helped stretch supply chains not only across countries and regions but also, in some cases, between continents. Globalisation requires Europe to be fully integrated into international transportation networks and has, for example, bolstered the role of ports like Rotterdam, Antwerp and Hamburg which can handle the largest vessels and air cargo hubs in Amsterdam Schipol, Frankfurt, London and Paris, for example. The long-term success of these international hubs depends on their onward connection with other transport modes, including road, rail and inland waterways to take goods into and out of Europe's economic heartlands.

In order to capture the interface between public policy and corporate strategy in the transport sector, it is helpful to discuss how the role and focus of key stakeholders have changed in recent years. In brief, there has been a shift in the balance of power between stakeholders: public stakeholders have relinquished some, but far from all, of their control. At the supranational level, the emphasis of the European Union is on a passive approach to create the conditions in which transnational transport infrastructure and services can flourish. The nation state continues to take responsibility for transport infrastructure construction, albeit less so than in the past, and, given the importance of transport to economic well-being, will always retain a strategic interest in it. Public actors at both state and regional level also promote transport infrastructure to achieve regional development objectives. The private sector's role in both transport infrastructure and services has become more extensive given the liberalisation and market access that has affected different transport modes and appears to be the main beneficiary of regime change. However, given the financial burden and commercial risks attached to some infrastructure projects, the public sector retains an operational role in most sectors to varying degrees.

The stakeholders in question are:

- Public stakeholders: the European Union;
- Public stakeholders: the nation state;
- Public stakeholders: regional and local authorities;
- Owners and operators of transport infrastructures;
- Transport and logistics companies;
- Customers and citizens.

Public stakeholders: the European Union

Although it is the European Union that is driving the TENs project, its room for manoeuvre in developing integrated trans-European infrastructure is limited. The main drivers behind transport infrastructure development in Europe remain the nation state, either directly or indirectly through planning regulations and the granting of concessions to private sector operators. The best the EU can do in creating the transport systems needed to service the needs of the single market and of globalisation is to persuade member states of the mutual benefits of co-operation and that planning should extend beyond and not stop at borders. As such, the most positive actions the EU can take in terms of driving TENs forward is to develop a framework to facilitate their emergence: harmonisation, standardisation and other measures to promote interoperability, support for research and development into satellite navigation[14] and the technology of traffic management all contribute to this goal. The Union has also developed the Marco Polo programme to facilitate intermodalism. Some funding comes from the European institutions for transport TENs but the total is relatively small given the overall size of the investment needed. Despite efforts to encourage private involvement, much of the funding for European transport infrastructure still comes from the national public purse.

The most striking initiatives the EU has taken in terms of transforming Europe's transport system are its efforts to liberalise the various modes of transport in Europe (see above). Prior to this, transport was amongst the most highly regulated and protected sectors within Europe: any competition that did exist was restricted to competition among operators of the same nationality. Securing access of Community transport operators to the transport networks of all member states was an important step in opening up trade routes and contributing to the attainment of SEM goals. However, despite the liberalisation element of the Common Transport Policy, it is difficult to argue that its enactment represents a complete shift from control to liberalism given the range of objectives of the EU's transport policy, including safety, the environment and, particularly, economic and social cohesion and improvement of the connectivity of peripheral regions.

Public stakeholders: the nation state

Planning and investment decisions regarding infrastructure have long been and remain a matter for the nation state (Short and Kopp, 2005; van Exel *et al*, 2002). Underpinning these decisions have long been issues of strategic and political concern: the routeing of roads, for example, was determined to a large degree by military factors in Roman times and for centuries afterwards the geography of the road network reflected the Roman legacy. The military legacy is also present for airports, many of which are built on former military airfields.[15] Airports and airlines have also long been regarded as symbols of national prestige, although this appears to be declining somewhat in Europe as the logic of liberalisation takes hold.[16]

The control aspect of transport policy also stretches into the highly politicised nature of decision making. The battle to influence the final outcome of infrastructure planning and transport policy decisions is strongly fought by a wide range of stakeholders and interest groups (including commercial interests such as construction companies and transport operators; environmental pressure groups, local groups with a variety of interests and differing regions competing for limited resources). The outcome of this political process is thus not necessarily that which would occur if the shaping of the transport sector was essentially left to the market.

With the pressures of European integration and globalisation, nation states are increasingly faced with the need for co-ordination of infrastructure planning with other countries. This requires assessment of the gains arising from the whole network and from all users, not just the portion of the network contained within a nation's borders or the benefits accruing to nationals. This can be difficult but is not impossible. The Øresund project, one of the completed Essen projects, is a good example of a successful collaboration between the governments of Sweden and Denmark and of the benefits to be gained in co-operating in cross-border infrastructure projects. This fixed road/rail link between Copenhagen and Malmö which opened in 2000 has helped create a new economic region with a common labour market with international business, particularly but not only in the pharmaceutical and biotechnology industries, choosing to locate in the region as their Nordic hub. The project also has supra-regional effects in terms of linking the Nordic region with the rest of Europe.

Not all cross-border projects are so successful and harmonious. The Channel Tunnel, although not part of the TENs initiative, which it predated, is a prime example of a major infrastructure project with dual nationality: the decision to build the Tunnel and the dual operation of it was enshrined in a treaty between the two countries. The upwards spiral of construction costs and lower than expected revenues[17] has left Eurotunnel mired in debt. As part of a rescue package, in May 2006

Eurotunnel plans to abandon its Anglo-French nature and restructure itself as a French parent company with a British subsidiary. Initial British parliamentary reaction to the plan is hostile on the grounds of the strategic importance of the project and the possibility has been raised that Eurotunnel's plans could be blocked if the changes are adjudged to breach its original concession to operate the tunnel.[18]

Public stakeholders: the regions and local authorities

Public stakeholders are also active at the sub-national level, both regionally and municipally. Regional airports are often owned by regional or local authorities and are seen as an important factor in both attracting inward investment and in connecting existing businesses to the outside world. Many distribution centres, for example, are located within easy reach of airports. Many seaports are also owned by local authorities: in the UK, for example, in 2006 65 ports were owned by local authorities. Although the traffic through these ports is generally not large in relative terms, it tends to be important to the regional hinterland and the ports themselves often act as important feeders to national or international ports. In a fully competitive situation, some of these ports would almost certainly go out of business but they are kept open for political and strategic reasons.

In addition to direct ownership and management of infrastructure, sub-national public authorities also often play an important role in the development of infrastructure by lobbying for investment to take part on its territory. This lobbying may or may not succeed depending on a wide range of political[19] and regional development reasons.

Owners and operators of infrastructure

Ownership patterns of infrastructure vary throughout Europe: roads, for example, tend to be owned and managed by the public sector in northern Europe whereas private sector involvement is much greater in southern Europe. Some countries, like the UK have privatised their airports and some of their seaports: elsewhere these nodes remain in public hands. In all cases, public authorities retain an element of control as illustrated by the examples below.

An interesting development in the mid-2000s has been the recognition by the private sector of the value of transport and infrastructure groups. This has manifested itself in a flurry of cross-border activities in Europe, the implications of which spread far beyond Europe's borders. In the first half of 2006 alone, initiatives included the controversial purchase by Dubai Ports World of P&O in 2006, a Goldman Sachs consortium led bid for Associated British Ports, and the acquisitions by two Spanish companies, Ferrovial and Abertis, of BAA, the British-based airports company and Italy's Autostrade respectively.

Ferrovial and Abertis are two examples of what could become a significant trend. Both are Spanish companies with roots in the construction industry. Both have broken out from this cyclically vulnerable sector, first into the construction and operation of toll roads and latterly into airports and related infrastructure. In 1998, 90 per cent of Ferrovial's pre-tax profit came from construction and real estate development. By 2005, the equivalent figure for infrastructure and services was 63 per cent. Prior to the BAA deal, Ferrovial had investments in 21 toll roads in Spain, Portugal, Ireland, Italy, Chile, Canada and the US, sizeable stakes in Bristol and Belfast City Airports plus management contracts in Chile and Australia plus major car parking investments in Spain. The acquisition of BAA, the world's operator of airports which owns seven airport in the UK, including the big three of Heathrow, Gatwick and Stansted, acquired the major shareholding in Budapest Airport in December 2005 and has operation management contracts in the US, Australia and Italy, pushes Ferrovial further away from the status of a construction company into the owner and operator of infrastructure in more than one mode.

Abertis has taken a similar path to Ferrovial. The company came into being as recently as 2003 following the merger of Acesa Infraestructuras and Aurea Concesiones de Infraestructuras, two companies involved in the construction, management and operation of the first Spanish toll motorways. Within two years of its formation, Abertis had become involved in telecommunications infrastructure and had acquired a controlling stake in the British airports operator TDI which operates eight airports, either under concession or as owner, four of which are in Europe (the UK and Sweden) and the others in the US and Bolivia. Abertis also manages five other airports in the US, Canada and Costa Rica. Like Ferrovial, Abertis has also extended its motorway activities beyond its domestic market and has interests of varying degrees in Portugal, France, Germany and the UK. Its biggest initiative to date has been the 2006 merger/acquisition of Italy's Autostrade which operates half of Italy's motorways and the whole of Italy's electronic tolling system.

At the time of writing (summer 2006), the Abertis-Autostrade deal still has to win the approval of the Italian government. The Italian Infrastructure Minister, Antonio Di Pietro, is reported to have expressed concerns about 'unacceptable violations' of the agreement made between the government and Autostrade when the toll concessions were granted. The eventual outcome will represent a benchmark of the extent to which the Italian government follows the logic of privatisation and leaves the operation to the market or whether it still seeks to exercise control, albeit at arm's length.

The emergence of infrastructure groups with interests in more than one transport mode and in several European countries (and beyond Europe's boundaries) is at an early stage. However, it is a trend that could well continue. The examples elaborated above have emerged from the motorway sector and, significantly from a country in which private operation of motorways was more advanced than in most of the rest of Europe. Further development of this trend offers interesting possibilities, particularly in terms of the potential for furthering intermodalism. However, even though motorways in parts of Europe are operated by the private sector and more airports are moving into the private sector, governments continue to exercise significant control through planning and regulation.

Transport and logistics companies

Increasingly, transport is regarded as part of a larger logistics function which is about all the costs of moving goods around and not just the cost of moving them from point A to B. Total logistics costs (transport, warehousing, inventory holding) have been estimated at 10–15 per cent of final products costs and total European spending on logistics at $1.2 trillion (CEC, 2006b). As transport and supply chains grow more complex, more and more European manufacturers and shippers are looking to outsource the logistics function to third party logistics suppliers (3PLs).[20] 3PLs plan and manage all warehousing and transport activities, including inventory management, tracking and tracing, supply chain management and, in some cases, secondary assembling of products. The aim of these companies is to provide an integrated service which links together all logistics functions and maximises the use of the transport system by exploiting intermodality where possible and appropriate.

Trends driving these developments include:

- Restructuring of European production structures as a result of market integration and enlargement, leading to a focus on a central, or at most limited number of, production sites;
- Reduced product life cycles and growing customisation of products, both of which require smaller and more reliable consignments;

- The impact of globalisation: low cost and outsourced production (perhaps in the Far East) and/or production chains that are integrated across national borders necessitate shipments over long distances and imply greater complexity. More companies are becoming multinational shippers and, as such, are seeking logistics suppliers that can supply services from the point of production to the point of sale;
- Technology changes, including the development of improved tracking and tracing systems.

An ideal 3PL operator is able to combine air, sea and road freight capabilities within an integrated network and offers their clients a 'one-stop shop'. Carbone and Stone (2005) argue that growth among 3PLs is aimed at providing more cohesive European logistics cover but that their efforts are constrained by national culture. Their efforts have also been severely constrained by the limited scope and indeed absence of pan-European logistics providers. Consequently, the first half of the 2000s has seen significant consolidation of the transport and logistics industry as companies seek to extend their geographical and modal scope.

As shown in Table 3.5, the route to becoming a pan-European logistics company varies tremendously, depending on the original starting point of the company. The trends demonstrate the strength of competition in the sector: although the liberal approach may not apply in all sectors (for example, the rail sector remains a long way from liberalism), the developing physical integration between sectors and vertical integration brought about by corporate restructuring has made the competition in this sector relatively intense. Firms can gain some relief from the competition by offering specialised services or new technologies but the overriding aspect of the industry is liberal.

Table 3.5 Mini-cases – emergence of pan-European logistics companies

The following mini-cases provide examples of the different origins of some of the major players in Europe's transport and logistics markets. These examples also demonstrate the new business opportunities opened out by increased market access and the pressure to take advantage of the opportunities created by more open markets – that is, by a greater liberalism generally. Moreover, by acquiring assets in various parts of the value chain, many companies are embarking upon a process of vertical integration. In the case of energy (see Chapter 5), this was harmful to competition. However, the reverse is opposite in Europe's transport and logistics industry as it brings companies into competition with each other that have, in the past, been located in different parts of the industry. Despite the convergence that is taking place, many companies remain distinctive because of strengths in specific parts of the industry.

Table 3.5 Mini-cases – emergence of pan-European logistics companies – *continued*

Rail – Deutsche Bahn:
Deutsche Bahn (DB) is unusual in that it has grown into a major multinational and logistics company from roots in the rail sector. DB came into existence in 1994 with the post-unification amalgamation of the Bundesbahn and the Reichsbahn. Since then it has followed a strategy of acquisition and diversification to become Europe's biggest rail freight operator, Europe's number one in land transport, third in global sea freight and fifth in global air freight with a total turnover of €24 billion in 2004. It has achieved this through the acquisition of logistics companies Stinnes and Schenker in 2002. These acquisitions not only gave DB, as the holding company, coverage of over 30 European countries but also a global reach. In 2006, DB completed the acquisition of US logistics company, Bax Global, which will extend the company's international reach. The aim is to create integrated logistics centres located at key intermodal freight hubs which provide the full range of logistics services. As a result of these changes, in 2004 34 per cent of DB's turnover was outside Germany and 46 per cent of it was from non-rail business. Moreover, over half of the business of Railion, DB's rail freight business, consists of cross-border traffic. All this has been achieved whilst DB has been a state-owned company. However, other freight and passenger companies have been granted access to the German rail network, which DB owns, and partial privatisation, possibly up to 25 per cent, is on the agenda for Angela Merkel's government.

Post offices: German and Dutch post offices
Liberalisation has been one of the driving factors behind changing strategies and infrastructure and infrastructure-related businesses. This has been particularly true for postal services which have been the starting point for breaking out into broader transport and related activities for the German and Dutch post offices, among others.

Until the late 1990s, Deutsche Post (DP) was essentially a state-owned domestic postal service. In 1998, 77 per cent of the company's revenue was from mail services. In 2000, it was privatised. Even before then, in 1999, DP acquired the Swiss logistics company, Danzas, at the time a world leader in air freight, the world's number two in ocean freight and with significant involvement in overland transport and supply chain management. By 2002, DHL, the international courier and parcels service founded in the US in 1969, become a DP company. This was followed by Exel, a UK-based transport and logistics company in 2005. Company revenues had tripled between 1998 and 2005: slow growth in mail revenues saw DP's share of total revenues fall to 28 per cent whereas the combined revenues of its Express and Logistics divisions reached 56 per cent of total revenues. In the space of a few years, DP had transformed itself from essentially a postal service into a multinational freight and logistics company.

The Dutch post office became a private company in 1989 and by 1995, the Dutch state had sold off sufficient of its shares to surrender its controlling interest. In 1996, the Dutch post office made a bid for international transport and logistics company, TNT. Since then TNT NV has become the holding company of the two brands – TNT and Royal TPG post. Through

Table 3.5 Mini-cases – emergence of pan-European logistics companies – *continued*

acquisitions and alliances, the group has built up its activities in mail, express and logistics. In the logistics area, it has developed expertise in inbound and outbound automotive logistics; consumer goods and retailing; high technology, tyres and publishing and media.

Consolidation of transport services and logistics providers
In addition to transport and logistics companies that have emerged from newly liberalised sectors, there are a heterogeneous group of private sector transport and logistics suppliers seeking to expand into full service provision across a wide geographical area whereas others are attempting to develop competences based on particular market segments such as refrigeration for fresh food, flowers and pharmaceuticals; solid bulk, primarily for agricultural and food products and liquid bulk for hydrocarbons and chemicals. Many of these companies have, like DB and the Dutch and German post offices, undertaken acquisitions to increase their geographical and modal coverage to consolidate their activities and to develop the critical mass needed to support pan-European operations. Some are family-owned firms like Germany's Dachser. Others like Gefco began life as the transport arm of a major manufacturer (in this case Peugeot) and have subsequently developed into an independent road transport operator and logistics integrator. UK-based Christian Salvesen's origins lie in shipping but by 1990 it had moved out of shipping and into industrial distribution and logistics. Subsequent acquisitions and dispersals have created a focus on seven European countries and logistics and strategic management for the outsourced supply chains for food, consumer and industrial products. DFDS Transport[21] chose to increase its market position by completing the acquisition of the Dutch company Frans Maas in 2006. This increased the company's volume, a necessary development in creating a European network. The company intends to consolidate its road network, to expand its air and sea networks and to develop its logistics capacity, an area which has been a particular strength of Frans Maas.

Customers and citizens

In a purely market-led world, the users of transport, both business and individual citizens, are key stakeholders. Indeed, businesses that are the users of transport are increasingly outsourcing their logistics function and, in the process, promoting serious competition among providers of services to fulfil their needs. Moreover, businesses have been quick to press governments for improvements in transport networks because of the cost of congestion for their activities. Indeed, it was the European Roundtable of Industrialists, an organisation of leaders of major European corporations, that was instrumental in getting the TENs initiative off the ground because of the adverse effect of inadequate infrastructure on Europe's competitiveness (see Chapter 2).

Similarly, the reactions of individuals to changes in transport markets are essential to the development of Europe's transport systems. The rise of budget airlines, which would have been impossible without the liberalisation of airline services that took hold in the 1990s, is a good example of how market changes can change passenger behaviour. However, individual consumers can also be highly resistant to modal shift and change. Private cars are one of the major contributors to road congestion, a problem which is almost universally acknowledged. However, drivers throughout Europe remain committed to retaining their cars and it is probable that fuel prices will have to increase much more and trains, the major competitor to cars on medium to long distance journeys, will have to increase the quality and frequency of their service significantly before drivers are prepared to give up the perceived advantages of driving.

Conclusion

As Table 3.6 shows, regime change has hit transport differentially, both in terms of modes and geographically. Before the launch of the single European market, transport in all modes was one of the most controlled and regulated sectors within Europe. The process of liberalisation and market access has, to a degree, been at work in all transport modes but their impact has varied, often depending on geography, timing and historically specific factors. It is, for example, no coincidence that the emergence of multi-modal European transport infrastructure companies has begun in southern Europe given the tradition of privately operated toll roads compared to the pre-dominance of almost total public sector control in northern Europe. Moreover, the emerging multi-modal logistics companies stem from and are shaped to an extent by their starting point, whether as liberalised postal services, as shipping companies or as traditional road hauliers, for example.

An ideal European transport network would be truly multi-modal with ample interconnections between modes and between national networks. Europe's transport system is being pushed slowly in that direction. In terms of the infrastructure, it is still the state at its various levels that retains a dominant role, particularly in terms of links such as road and rail. However, there has been more easing of control in relation to some nodes with the growing commercialisation of airports and seaports. Even in the latter cases, the state retains an intense interest because of regional development and other strategic issues. In terms of service provision, however, the market has become a more powerful driver. This is particularly noticeable, not so much in any one mode

Table 3.6 Summary of frameworks in which transport sectors operate

Roads – Infrastructure	*Southern Europe:* tendency to corporate – emergence of multinational motorway operators (possible strategic backlash following Abertis-Autostrade deal) *Central and northern Europe:* control remains dominant.
Road – services	Haulage – a liberalised sector, ranging from one-man operations to specialised and pan-European hauliers and integrated into multinational logistics providers.
Civil Aviation – infrastructure	Control – many airports remain in public hands, either at national, regional or local level. Often seen as an important factor in regional development. Strategic – many privatised airports, often subject to competition and to active public policy (e.g. planning authorisations and regional policy initiatives).
Civil Aviation – services	Liberal – emergence of low cost carriers has enhanced competition on some routes/market segments. Corporate – development of strategic alliances limits competition and improves efficiency.
Rail – Infrastructure & services	Largely control – still mostly state monopolies (limited exceptions). Modal shift aspirations imply continuation of active policy plus continuing subsidisation of many services. Liberalisation – early stage at European level but policy objectives placed on rail (environmental, etc.) implies active policy and strategic approach. Already liberalised national rail (e.g. UK) – continuing strategic approach because of conditions of concessions/regulatory regime, etc.
Seaports	Mixture of control and strategy – ownership of ports is mixed (private and public – national, regional and local) but regional development role of ports tends to encourage public involvement to some degree. Growing element of competition between ports and between regions.
Inland waterways – Infrastructure	Control – removal of bottlenecks carried out by the public sector.
Services	Corporate/liberal – mostly private operators but obstacles to smoothly operating market. NAIADES programme – a passive approach intended to provide a framework which will facilitate competition.

(although airline competition has become particularly intense) but in the way in which operators seek to gain a competitive advantage and in the way in which they provide their services across modes. As such, increased pressure on infrastructure providers by service providers and users can be expected to contribute to the objective of trans-European transport networks.

Given the nature of transport infrastructure (that is, high cost, high risk, longevity and the persistence of control – which in part continues because of the preceding three factors), the watchword in its development is more co-operation than competition. Conversely, in relation to transport services, the regime change towards greater competition has been, on the whole, much more marked.

References

Anguera, R. (2006) 'The Channel Tunnel: an ex-post economic evaluation', *Transportation Research*, Part A 40(2006), pp.291–315.

Carbone, V. and Stone, M. (2005) 'Growth and relational strategies used by the European logistics service providers: rationale and outcomes', *Transportation Research,* Part E 41, pp.495–510.

Commission of the European Communities (CEC) (1960) *Third General Report on the Activities of the Community.*

Commission of the European Communities (CEC) (1979) *The role of the Community in the development of Transport Infrastructure,* Bulletin Supplement 8/79.

Commission of the European Communities (CEC) (1985) *Completing the Internal Market,* COM(85) 310.

Commission of the European Communities (CEC) (1992) *Commission Communication and Proposal for a Council Decision on the Creation of a European Inland Waterway Network,* COM(92) 231.

Commission of the European Communities (CEC) (1994) *Proposal for a European Parliament and Council decision on Community guidelines for the development of a trans-European transport network,* COM(94) 106.

Commission of the European Communities (CEC) (1995) *Amended proposal for a European Parliament and Council Decision on the Community guidelines for the development of the trans-European transport network,* COM(95) 48.

Commission of the European Communities (CEC) (2001) *White Paper – European transport policy for 2010: time to decide,* COM(2001) 370.

Commission of the European Communities (CEC) (2004a) *Commission Staff Working Paper: Trans-European Transport Network – Report from the Commission to the European Parliament, the Council, the European Economic and Social Committee and the Committee of Regions on the Implementation of the Guidelines for the Period 1998–2001,* SEC(2004) 220 of 19 February 2004.

Commission of the European Communities (CEC) (2004b) *Proposal for a Directive of the European Parliament and of the Council amending Council Directive 91/440/EC on the development of the Community's Railways,* COM(2004) 139 final of 3 March 2004.

Commission of the European Communities (CEC) (2004c) *Further integration of the European rail system: third railway package,* COM(2004) 140 final of 3 March 2004.

Commission of the European Communities (CEC) (2004d) *Proposal for a Directive of the European Parliament and of the Council on the certification of train crews operating locomotives and trains on the Community's rail network,* COM(2004) 142 final of 3 March 2004.

Commission of the European Communities (CEC) (2004e) *Proposal for a Regulation of the European Parliament and of the Council on International Rail Passengers' Rights and Obligations* COM(2004) 143 final of 3 March 2004.

Commission of the European Communities (CEC) (2004f) *Proposal for a Regulation of the European Parliament and of the Council on compensation in cases of non-compliance with contractual quality requirements for rail freight services,* COM(2004) 144 final of 3 March 2004.

Commission of the European Communities (CEC) (2004g) Commission Staff Working Paper: *Proposal for a Directive of the European Parliament and of the Council amending Council Directive 91/440/EC on the development of the Community's Railways to gradually open up the market for international passenger services by rail – Extended Impact Assessment,* SEC(2004) 263 of 3 March 2004.

Commission of the European Communities (CEC) (2004h) *Energy and Transport in Figures – 2004,* http://europa.eu.int/comm/dgs/energy_transport/figures/pocketbook/doc/2004/pb2004.pdf – accessed November 2005.

Commission of the European Communities (CEC) (2005a) *Draft Commission Decision designating six European coordinators for certain trans-European transport network,* July 2005, http://ec.europa.eu/ten/transport/agency/doc/2005_07_02_coordinator_en.pdf accessed October 2005.

Commission of the European Communities (CEC) (2005b) *Trans-European Transport Network: TEN-T priority axes and projects 2005,* Brussels: DG Tren.

Commission of the European Communities (CEC) (2005c) *Energy and Transport in Figures – 2005,* http://ec.europa.eu/dgs/energy_transport/figures/pocketbook/doc/2005/etif_2005_transport_en.pdf – accessed May 2005.

Commission of the European Communities (CEC) (2006a) *Communication from the Commission on the Promotion of Inland Waterway Transport 'NAIADES' – An Integrated European Action Programme for Inland Waterway Transport,* COM(2006) 6 final.

Commission of the European Communities (CEC) (2006b) *Logistics for Promoting Freight Intermodality,* presentation by Ismo Koskinen at Consultation Workshop, Brussels, 25 April 2006.

Community of European Railways (CER) (1989) *Proposals for a European High-Speed Network,* Brussels: CER.

Eurostat (2004) *European Business: Facts and Figures – 1998–2002.*

Exel, J. van, Reinsure, S., Gommers, M., Pearman, A. and Tsamboulas, D. (2002) 'EU involvement in TEN development: network effects and value-added', *Transport Policy,* 9(2002): 299–311.

High Level Group (2003) *Report on the Trans-European Transport Network* (also known as *the Van-Miert Report*), http://ec.europa.eu/ten/transport/revision/hlg/2003_report_kvm_en.pdf.

Official Journal (1978) *Council Decision of 20 February 1978 instituting a consultation procedure and setting up a committee in the field of transport infrastructure,* OJC 54 (78/174/EC).

Official Journal (1991) *Council Directive of 29 July 1991 on the development of the Community's Railways (91/440/EC)*, OJL 237 of 24 August 1991, pp.25–8.

Official Journal (1996) *Council Directive 96/48/EC of 23 July 1996 on the interoperability of the trans-European high-speed rail system*, OJL 235 of 17 September 1996, pp.6–24.

Official Journal (2001a) *Directive 2001/12 of the European Parliament of 26 February 2001 amending Council Directive 91/440/E on the development of the Community's railways*, OJL 75 of 15 March 2001, pp.1–25.

Official Journal (2001b) *Directive 2001/12 of the European Parliament of 26 February 2001 amending Council Directive 95/18/EC on the licensing of railway undertakings*, OJL 75 of 15 March 2001, pp.26–8.

Official Journal (2001c) *Directive 2001/12 of the European Parliament of 26 February 2001 on the allocation of railway infrastructure capacity and the levying of charges for the use of railway infrastructure and certification*, OJL 75 of 15 March 2001, pp.29–46.

Official Journal (2001d) *Directive 2001/16/EC of the European Parliament and of the European Council of 19 March 2001 on the interoperability of the trans-European conventional rail system*, OJL 110 of 20 April 2001, pp.1–27.

Official Journal (2002) *Commission Decisions 730–735 of 30 May 2002* – high speed TSIs, OJL 245 of 12 September 2002, pp.1–506.

Official Journal (2004a) *Corrigendum to Decision No 884/2004/EC of the European Parliament and of the Council of 29 April 2004 amending Decision No 169/96/EC on Community guidelines for the development of the trans-European transport network*, OJL 201, pp.1–55.

Official Journal (2004b) *Regulation (EC) No 881/2004 of the European Parliament and of the European Council of 29 April 2004 establishing a European Railway Agency*, OJL 164 of 30 April 2004. A corrigendum has been published in OJL 220 of 21 June 2004, pp.3–15.

Official Journal (2004c) *Directive 2004/50/EC of the European Parliament and of the European Council of 29 April 2004 amending Council Directive 96/48/EC on the interoperability of the trans-European high-speed rail system and Directive 2001/16/EC of the European Parliament and of the Council on the interoperability of the conventional rail system*, OJL 164 of 30 April 2004. A corrigendum has been published in OJL 220 of 21 June 2004, pp.40–57.

Official Journal (2004d) *Directive 2004/51/EC of the European Parliament and of the European Council of 29 April 2004 amending Council Directive 91/440/EC on the development of the Community's railways*, OJL 164 of 30 April 2004. A corrigendum has been published in OJL 220 of 21 June 2004, pp.58–60.

Official Journal (2004e) *Directive 2004/49/EC of the European Parliament and of the European Council of 29 April 2004 on safety on the Community's railways and amending Council Directive 95/18/EC on the licensing of railway undertakings and Directive 2001/14/EC on the allocation of railway infrastructure capacity and the levying f charges for the use of railway infrastructure and safety certification*, OJL 164 of 30 April 2004. A corrigendum has been published in OJL 220 of 21 June 2004, pp.16–39.

Official Journal (2006a) *Commission Regulation (EC) No 62/2006 of 23 December 2005 concerning the technical specification for interoperability relating to the telematic applications for freight subsystems of the trans-European conventional rail system*, OJL 13 of 18 January 2006, pp.1–72.

Official Journal (2006b) *Commission Regulation (EC) No 66/2006 of 23 December 2005 concerning the technical specification for interoperability relating to the subsystem rolling stock – noise of the trans-European conventional rail system,* OJL 37 of 8 February 2006, pp.1–49.

Short, J. and Kopp. A. (2005) 'Transport infrastructure: investment and planning – policy and research aspects', *Transport Policy,* 12(2005): 360–7.

Vinois, Jean-Arnold (2002) *Creation of a European railway area against the background of the White Paper on European Transport Policy,* http://europa.eu.int/comm/transport/rail/overview/doc/ri-5jav-en.pdf – accessed December 2005.

4

Trans-European Telecommunication Networks

The shift towards a liberal regime within European telecommunications is tied intimately into the development of the information society across the continent. The rise of market forces across telecommunications (as well as increasingly related sectors such as IT and content) has created an interface between corporate strategy and public policy that is driving widespread socioeconomic change. The Commission's aim is to use the public policy/corporate strategy interface to build the information society based on a pluralist platform reflecting its liberal-driven development agenda. However this pluralist strategy understated the aggressiveness (in both political and economic terms) with which incumbents sought to defend their respective positions. As a result, doubts emerged over the viability and practicality of the liberal approach. This concern has been compounded by the slump in telecommunications in the late 1990s and the uncertainties and instability within corporate strategy that it created. Thus the development of TENs in telecoms stands at a crossroads between the desire for a market-driven approach and the need to utilise incumbents to deliver the information society. This will involve a need to control and reward these incumbents which has implications for the policy/corporate strategy interface. This paradox at the heart of policy/strategy development will be the key theme of this chapter.

Initially this chapter looks at the EU policy actions to promote the information society through liberalism. Consequently, the chapter examines the evolutionary shift towards a liberal agenda for telecommunications culminating in the Electronic Communications Framework Directive. Thereafter the chapter moves to examine the key market-driven responses to these opportunities. From this analysis, key issues in the development of the information society are addressed, notably issues regarding access to the spread of broadband, the local loop and emergent mobile strategies. Before conclusions are reached, the policy measures within both a national and supranational context are examined.

The European information society

The information society has moved to the centre of the EU's international competitiveness policy through seeking to stimulate the broad socioeconomic changes associated with the increased strategic importance of information and knowledge (for a review, see Garnham, 1997). The European Commission's path for the development of the information society is intimately linked into the re-regulation of telecommunications (and other related sectors) and the spread of advanced (most notably broadband) infrastructure. Through this strategy, the EU has sought to establish a clear link between the development of the information society, the corporate strategies of telecommunications operators and the international competitiveness of the EU.

The links between telecommunications policy and the creation of the information society only really started to become an explicit policy goal in the 1990s (see below). In doing so, the EU promoted a version of the information society that was more socially inclusive than the form promoted by the US which was more explicitly market-driven (especially in terms of deliverable content). In tune with the generic shift towards liberalism within the global economy, the EU recognised that such changes could only be realised through the freedom for operators to create and deploy the networks that act as the necessary platforms for the evolution of the information society (Liikanen, 2002). Consequently, the EU sought to remove the key impediments to such freedoms not least of which were the powerful telecommunications incumbents (see Table 4.1) who have a vested interest in limiting market integration. Core to the emergence of the European information society is facilitating the free mobility of information. Thus, just as the SEM has sought the realisation of the four freedoms (labour, capital, goods and services), so there is a need to create a fifth freedom – information. This requires integrated networks and a pan-European service environment (Natalicchi, 2001).

From the mid-1980s, the Commission (in concert with actions by national authorities) set about challenging the paradigm of network development based on control and monopoly. The European Com-

Table 4.1 State ownership and market power of incumbents (September 2005)

State (Incumbent)	State share (%) (Sept 2005)	Market share of mobile subsidiary (%) (Oct 2005)	Market share of fixed telephony % (Dec 2004) – (by retail revenue)	Fixed broadband retail lines market share % (July 2004)
Belgium (Belgacom)	50 plus one share	47	53	51
Czech Republic (Cesky Telecom)	0	n/a	n/a	39
Denmark (TeleDanmark)	0	31	61	64
Germany (Deutsche Telekom)	38	38	57	87
Estonia (Elion Ettevotted)	27	46	84	50
Greece (OTE)	36	n/a	76	40
Spain (Telefonica)	0	48	79	56
France (France Telecom)	35 (approx.)	47	72	48
Eire (Eircom)	0	0	79	69
Italy (Telecom Italia)	0	40	86	70
Cyprus (Cyta)	100	93	97	100
Latvia (Lattelekom)	51	0	93	81
Lithuania (Lietuvos Telekomas)	2	0	93	34
Luxembourg (P&T Luxembourg)	100	58	n/a	74
Hungary (Matav)	0 (though keeps a golden share)	45	96	50
Malta (Maltacom)	60	48	100	33
Netherlands (KPN)	21 (plus a golden share)	36	65	44
Austria (Austria Telekom)	30	40	55	35
Poland (Telecominikacja Polska)	4	34	83	94
Portugal (PT)	7 (plus one golden share)	n/a	84	79
Slovenia (Telekom Slovenije)	72	74	98	71
Slovakia (Slovak Telecom)	49	44	98	46
Finland (TeleSonera)	14	n/a	n/a	72
Sweden (TeleSonera)	45	n/a	n/a	40
UK (BT)	0	0	51	25

mission (and many governments) believed that incumbents had neither the pressure, desire nor ability to lead Europe in a coherent manner towards the information society. However the Commission recognised that aggressive confrontation and a rush towards a pure market-driven process could result in market failure leading to aspects of the broad socioeconomic context of the information society remaining unfulfilled. As a key theme of the European information society was 'digital inclusiveness', the European Commission felt that a rush towards market-driven processes could reinforce existing differences between socioeconomic groups and regions creating a digital divide both within and across states. As such market failures needed to be avoided, the challenge to the prevailing development paradigm was to be more evolutionary than revolutionary. However, Jordana (2002) argues that throughout the 1990s, these broad market failure remedies played a backseat to the process of liberalisation.

Whilst liberalisation was a gradual process, it did focus incumbent minds as competition began to intensify. Many new entrants began to carve-out niches in market segments that were either the most profitable or where many incumbents did not – at the time – see much potential (most notably mobile). With confidence rising in the telecoms sector throughout the 1980s and 1990s, the Commission felt that the social applications (that were central to its vision of the information society) would emerge naturally from market forces. This was based on an expectation that competitive intensity would drive operators to offer new services as they sought to generate extra traffic. However, by the mid-1990s, this confidence in the liberal model seemed over optimistic as there was little evidence (at least over the short term) of such linkages. For many operators, social applications simply did not generate the traffic levels to warrant investment from private sources. Thus support from public bodies was still necessary (see below).

An emerging framework for the development of telecommunications infrastructure

Traditionally pan-European networks were based upon interconnection through national monopolies as major policies and investment plans were established by technical experts largely outside public scrutiny (Noam, 1992). Over time it was becoming evident that the delivery of the economic and social services/applications associated with the emergent information economy could and should not really be delivered through a single monopoly supplier. As a result, the pattern of network development needed a regime change based on a 'modular' network environment (Noam and De Sola, 1980). According to Noam (1992) this breakdown was caused by a number of factors, including:

- Technological changes that led to new transmission options that eroded the monopoly;
- The convergence of IT and telecommunications;
- A need to curtail any forces within the emergent information economy that could render a single firm too powerful;
- That government regulation was incapable of controlling a monopoly;
- That large businesses could successfully fight the monopoly system;
- That the diversification of telecommunications made it impossible for one provider to serve all submarkets;
- That monopoly inefficiency lead to the emergence of competitors;
- That competition is linked to socioeconomic progress;
- That the very success of the public network in creating universal service leads to the emergence of new networks.

These factors in combination created powerful forces to move the sector towards a more liberal regime. The catalyst for this regime shift has been through the progressive liberalisation of the telecommunications sector which has – according to Larouche (2000) – occurred through the following series of stages.

The starting model

The initial model of network development is based on monopoly supply where cross-border services were undertaken *via* the correspondent system. Under this system, services between states occurred through mutual co-operation. Thus networks are interconnected and interoperable through facilities for which the respective national operator was responsible. This meant that each operator took responsibility for the termination of traffic within their respective domain. The operator in the state where the call originated collected the revenue from the call and then gave a proportion to the terminating partner. At this stage, within each EU member state, there was respective national hegemony in the development of telecommunication systems *via* state ownership and monopoly control. Prior to the initial phase of liberalisation, EU action in telecoms networks was linked to standards development, shared research, aid for less developed regions and common stances in international actions. These policy themes have continued throughout the period of regulatory change.

The regulatory model (1990–96)

The 1987 Green Paper started to change this system. Whilst leaving infrastructure and public voice telephony (the cash cow for incumbents) under monopoly control, the Green Paper liberalised all other services. In addition, the paper sought community wide interoperability through harmonised standards and Open Network Provision to regulate the relationship between the monopoly infrastructure provider and competitive service providers. The liberalisation of value-added services facilitated market segmentation as new businesses emerged to offer new services such as mobile and advanced data products. However, new operators were still heavily dependent upon the incumbent for service delivery. The move towards the development of a common market for telecommunications was given increased political and commercial impetus through the SEM programme which acted as a basis for the further liberalisation of the sector.

Transitional model (1996–97)

In the mid-1990s, the Commission launched an ambitious timetable to seek the full liberalisation of the market. The aim was – by 1996 – to liberalise alternative infrastructure (all non-incumbent owned infrastructures) immediately and seek full liberalisation by 1998. The ambition was to gradually increase the choice of infrastructure for non-incumbent service providers and put pressure upon the remaining monopoly elements of the incumbent supplier. This policy strategy also reflected the sharp rise in network usage driven by the emergence of mobile telephony and a belief that existing regulatory frameworks were unable to support its rise.

The fully liberalised model (from 1998)

The final phase of liberalisation (from 1998) exposed the remaining aspects of monopoly provision (voice and infrastructure provision) to competition (see Waverman and Sirel, 1997). This shifted the regime of network development towards greater modularity, further expanding the choice of non-incumbent operators over transmission infrastructure. It also offered scope to remove restraints on network capacity and sought to increase the ability of infrastructure to cope with the expected increases in traffic. The impulse for these measures was driven by a broader political desire to restore the growth of the European economy by shifting it towards knowledge and information intensive activities. The Commission Growth Strategy (CEC, 1998) outlined the need for the development of a common information area replacing national fragmentation with a federal network of networks.

* * *

Within this phased approach to liberalisation, there were derogations for the EU's 'less developed' states especially with regard to infrastructure and voice liberalisation (Preston, 1995). In many cases, incumbents successfully pressed national policy bodies to negotiate for extra time to allow them to adjust to the new regime. In practice, such has been the rapidity of development of the telecommunications system that the need for extra time was often not required or could have been counter-productive (Berben and Clements, 1995). As competition has matured, so the source of value within the network has become increasingly focused on access to the end user. This shift has intensified disputes between new entrants and incumbents focused on access to the local loop (see below). Thus whilst non-incumbent operators have been developing pan-European infrastructures (see below), there have been limits upon their ability to develop an end-to end infrastructure as incumbents have sought to retain control of user access which is still under their control. Overall market opening on its own is insufficient to create meaningful competition in telecommunications due to:

- Strong network effects which lead users to prefer the larger networks;
- Large sunk costs especially in creating essential facilities;
- Long legacy of statutory monopoly that creates in-built user inertia;
- Sustained abusive behaviour by incumbents.

The result has been that governments and other public administrations still need to regulate. Thus there has been little desire to move direct to competition law without some sector specific regulation.

Despite these remaining roadblocks, the process of deregulation signalled the beginning of the end of the unitary network and the rise of multiple networks based upon competitive supply. As Mansell (1993) identified, there has been a shift towards a more idealistic framework based on liberal economics. However the development of the information society as a tool of public policy has been overtaken by commercial processes and by the strategic actions of corporate entities. The spread of the Internet, the sharp rise in mobile phone usage offered new platforms for the development of the information society that were not envisaged when the process was initiated. This was aided by the nature of the Internet as an open, non-proprietal system. It was only with the maturity of liberalisation that it became evident that these technologies would act as major platforms for the development of the information society.

The electronic communications framework

In 1999, the Commission launched a review of the telecommunications regulatory structure with the aim of bringing all communications infrastructures and services under a single framework. The aim was to remedy any fragmentation of national regulations and mark a path from sector specific regulation to the application of generic competition provisions. The policy sought to give national regulators greater discretion to deal with specific regulatory problems as well as developing a framework for more co-operation between themselves and the Commission (Berhin *et al*, 2005).

The new regulation was novel in that it attempted to establish a set of generic rules for ex-ante intervention as well as carrying forward existing regulations in areas such as universal services. As a result, not only was member state discretion in imposing individual licenses limited but the framework leant heavily on competition law to identify players with significant market power. This is reflected in the guide analysis on the size and measure of market power (establishing a common methodology for the National Regulatory Authorities (NRAs)) offered by the Commission. In addition, the framework offered recommendations in 18 submarkets where the NRAs have identified that competition could be imperfect. If significant market power (which is based on having 25 per cent market share) is found to exist, the NRAs are compelled to take action to remedy this problem (CEC, 2003). These policies reflect a general stance within the regulatory framework to reduce the administrative burden upon operators as well as offer a lighter regulatory approach and a recognition that competitive impediments at the local and national level are best dealt with in this domain.

The regulatory framework consists of the following directives which were meant to be implemented by July 2003:

- Framework Directive: outlines the general principles, objectives and procedures;
- Authorisation Directive: replaces individual licenses by general authorisations to provide communications services;

- Access and Interconnection Directive: sets out rules for a multi-carrier marketplace, ensuring access to networks and services, interoperability, and so on;
- Universal Service Directive: guarantees basic rights for consumers and minimum levels of availability and affordability;
- e-Privacy or Data Protection Directive: covers protection of privacy and personal data communicated over public networks.

As Table 4.2 indicates there is still someway to go before these directives are fully implemented across all states. As a result, the Commission has

Table 4.2 State of transposition of electronic communications directives (October 2004)

	Framework	Access	Authorisation	Universal service	e-privacy	Competition
Belgium	*	*	*	*	*	*
Czech Rep.	*	*	*	/	x	*
Denmark	/	/	/	/	/	/
Germany	/	/	/	/	/	/
Estonia	x	x	x	x	x	x
Greece	x	x	x	x	x	x
Spain	/	/	/	/	/	/
France	*	*	*	*	/	/
Ireland	/	/	/	/	/	/
Italy	/	/	/	/	/	/
Cyprus	/	/	/	/	/	/
Latvia	/	/	/	/	/	/
Lithuania	/	/	/	/	/	/
Luxembourg		x	x	x	x	x
Hungary	/	/	/	/	/	/
Malta	/	/	/	/	/	/
Netherlands	/	/	/	/	/	/
Austria	/	/	/	/	/	
Poland	/	/	/	/	/	/
Slovenia	/	/	/	/	/	/
Slovakia	/	/	/	/	/	/
Finland	/	/	/	/	/	/
Sweden	/	/	/	/	/	/
UK		/	/	/	/	/

/ – Full transposition
* – Partial transposition notified/partial information supplied
x – No transposition notified/no substantial information supplied

opened up infringement proceedings against some states (many of which are the newer members in Central and Eastern Europe) to ensure this position is rectified.

Overall, whilst regulation is seen as necessary, it is also perceived as ultimately temporary as a means of supporting new entrants over the short term. In addition, the new framework also specifies technological neutrality and consistency across the EU. These measures seek to ensure flexibility in the approach to deal with emerging technologies and to allow collaboration between NRAs. However there are concerns by the Commission that there is an insufficient gap between the regulator and the incumbent operator. This is an especially of concern where the state still has a substantial shareholding in the operator (see Table 4.2). Thus the independence of regulation has not been guaranteed. There have also been barriers by the state to stop NRAs from carrying out their duties and collecting the relevant information required (Daßler and Parker, 2004).

Market-driven processes

Deregulation combined with technological change created what Fransmann (2002) terms a consensual vision for telecommunications. In the mid-1990s, market-driven processes within the telecommunications promoted the following commonly held perceptions (what Fransmann terms the 'consensual view') as to the sector's development:

1. Many expected an explosive demand for bandwidth based on the spread of the Internet coupled with high price elasticity for network capacity. Thus, amongst operators, a belief existed that if they built the price infrastructure, users would come with the result that any reductions in price of network capacity would be more than compensated by the rise in traffic that resulted;
2. There was a consensus that the new operators would out-compete incumbents as they were better focused, more technologically adroit and possessed flatter organisational structures. The strategic focus of these businesses upon cream skimming the incumbent's most lucrative markets meant that the latter would be left with low growth legacy businesses;
3. There was an expectation that financial markets would support the fittest new entrants enabling these higher growth businesses to build, develop and utilise their own infrastructure networks;

4. There was a belief that technological change would reinforce these changes further undermining the position of the incumbents as the new entrants would be quicker to deploy new technologies and develop new services.

These beliefs contributed to the rapid growth in the telecommunications sector in the EU since 1998. Between 1999 and 2004, the market grew from €163 billion to €277 billion. As the market grew, so there was a sharp rise in new entrants which in the fixed line segment (in the EU-25) rose from 526 in 1998 to 2,141 in 2004 (CEC, 2004b). The increase in competitive intensity has been fastest in the more mature marketplaces (such as the Nordic states). Not surprisingly, this rise in competition has gone hand in hand with a reduction in the market power of incumbents whose share of the voice telephony market has fallen by as much as 20 per cent. In addition, the choice for consumers has risen with up to a third now using an alternative provider for some form of communications service (CEC, 2004b). Overall, the market share of incumbents has fallen in all market segments (except local calls) with international and mobile calls suffering the fastest erosion. Nevertheless this should not disguise the fact that incumbents still have high market share across the EU-25 (see Table 4.1) with the choice of suppliers in many member states still limited. Thus the re-regulation of the sector has created new niches for businesses which have tended to erode market positions of incumbents on a segment-by-segment basis. This erosion has been fastest in the higher growth segments with its dominance being sustained in more mature segments.

The other key development has been the growth of the mobile market which has risen in value from €98 billion in 2001 to over €122 billion in 2004. Mobile services now represent some 45 per cent of the EU telecoms market with traditional fixed voice comprising 33 per cent. Consequently, mobile subscribers have risen sharply as the penetration rate has risen from 18 per cent (for the EU-15) in 1998 to 92.8 per cent in 2005 (for the EU-25). In eight states, the penetration rate is now over 100 per cent. Subscriber numbers in the EU-15 have risen from 69 million in 1998 to 426 million in 2005. There has also been a substantial rise in the number of mobile providers. In 1999, the EU had just 52 authorised mobile suppliers. By 2005, this had risen to 166 with the UK alone having 59. This has been accompanied by reduction in market share of the mobile market by relevant incumbent subsidiaries with some incumbents (most notably BT) divesting their mobile assets (see Table 4.1). Overall, the average market share of mobile communications of the incumbent has fallen from 64 per cent in 1998 to 43.2 per cent in 2005.

The emergence of pan-European networks

These changes in the service environment have fed through into increases in investment in pan-European backbones. Pan-European backbones are extensive multinational networks that operate largely as a transmission function to facilitate the movement of large quantities of data over relatively long distances. On the back of expectations of rising traffic, pan-European infrastructure owners deployed a peak of 57,000 km of network in 2001; up from 3,600 route km in 1997 (FT, 2001a). Overall total European network capacity has risen from 339 gbp/s[1] in 2000 to around 9.2 tbp/s[2] per second in 2005 (FT, 2001a, b, c). Not only was more capacity being laid in terms of more fibre put into the ground but capacity was also being extended by technological change that increased the capacity of existing infrastructures manifold (FT, 2001b).

The development of pan-European backbones was through a mix of international expansion strategies by incumbents such as BT, new entry by US operators (such as Global Crossing and GTS) and by new alternative suppliers such as Colt. By 2001, there were 27 separate pan-European carriers. The main focus of this network build was the four leading financial centres of Amsterdam, London, Paris and Frankfurt where both MNCs and bandwidth hungry sectors (such as finance) were located and which generate over 60 per cent of all European traffic (Rutherford *et al*, 2004). Whilst the new entrants and incumbents have sought to compete in the development of pan-European networks, their strategies differ with some operating as wholesalers and others as retailers. Over time, the demand for international traffic capacity has spread to smaller business and other forms of cross-border traffic flows stimulated in no small part by the spread of the Internet and the rise of electronic commerce.

The methods or market entry in the development of networks were manifold:

1. Direct build: operators (such as Viatel) who built their own networks;
2. Co-building: sharing the cost of infrastructure building with others (such as Interoute, Colt and Level 3);

3. Indefeasible right of use (IRUs): carriers built networks based on long-term leased capacity which in effect means the operator can act like it owns the fibre. This is the quicker method of network development;
4. Leasing: based on short-term pay as you go usage capacity with the user generally paying a fixed amount per month;
5. Swaps: to mitigate the transaction costs involved in developing IRUs or leases, operators simply swap capacity with one another.

Carriers have used all these methods to create the reach and penetration desired to develop a network configuration across the EU that meets the demands and needs of business. BT, for example, built a European network by purchasing and integrating a dispersed set of national assets. Financial and time-to-market issues were clearly in the minds of operators as they chose their market entry strategy. Of the 27 firms that were developing pan-European networks, it is estimated that, by 2000, only five were actually building in the sense of deploying an actual infrastructure. Though a lot of fibre has been deployed not all has been 'lit' i.e. actually turned on to be capable of carrying information. Indeed it was estimated that by 2001, only 1–2 per cent of the capacity of European's infrastructure was actually ready for use (FT, 2001a). Thus the vast majority of Europe's network capacity is 'dark fibre'.[3] As a result, operators were caught in a quandary as to whether to deploy technologies (such as compression technology) that increased the capacity of their existing network or to utilise spare capacity.

As highlighted above, the development of these pan-European networks focused on those (mainly urban) locations where there is a high density of users and where a higher volume of traffic is likely to be generated. This highlights the uneven market-led development of the information society across Europe. These developments merely reinforced historical trends of infrastructure development as these networks were driven by large corporates. Research by Rutherford *et al* (2004) highlight that while there is over-investment in Northern Europe (as mentioned above) there is also substantive investment in 'gateway' cities. These gateway cities act as a funnel for traffic from the more peripheral regions of the EU. They conclude that major European backbones rely on a minimum of 12–15 cities to deliver high bandwidth networks and services across Europe. This is larger than pan-US networks and reflects continuing national territorialities. This underlines that despite the development of pan-European networks, a territorial mindset still seems important in network development.

The telecoms crisis in Europe

By 2004, less than a quarter of the 27 pan-European operators that existed in 2000 continued to trade. In part, this was a spillover from the US where many of the 23 operators (such as WorldCom and Level 3) that went bankrupt had substantial European operations. The causes of this crisis lay in over-investment in European telecommunications infrastructure due to the fact that many of the beliefs about the nature of the market created by the aforementioned consensual view of the sector trends proved to be misplaced.

First, the assumption of a high price elasticity of demand for network capacity and the consequent 'build it and they will come' attitude that resulted proved misplaced. Between 1997 and 2001, Internet traffic doubled every year. However much of the industry was working on the basis that it would double every 100 days. As a result the quantity of deployed fibre increased five fold between 1998 and 2001 with technological change increasing the capacity of each strand of fibre 100 fold (FT, 2001a). The result was a 500 fold increase in capacity whilst demand only quadrupled. Thus despite high price elasticity of demand, the overhang of capacity was such that – even though network usage still rose – it was less than any increase in capacity with the result that prices fell sharply pushing the core product of the telecommunications sector – network capacity – towards commodity status.

Second, operators based market entry strategies on the belief that competitive advantage was linked to the size of the network (*The Economist*, 2002). This assumption was based on Metcalfe's law which justified the building of large networks as it stated that the number of cross connections (hence the number of users, etc.) is proportional to the square of the number of nodes and users. The result is that bigger is more than proportionally better. However, this resulted in firms expanding capacity, network size and accruing large debt. As revenues failed to rise so the debt burden forced many operators into bankruptcy (FT, 2001a, b). As a sector, telecommunications networks incur high fixed and sunk costs and low marginal costs with the result that firms incur huge costs in developing networks but the cost of serving an extra customer is low.

Third, it was assumed that the time that the Internet would take to spread throughout the EU economy would be relatively short. This also proved misplaced as the spread of the Internet took longer to diffuse than previously supposed. As a result, the anticipated traffic levels expected to be generated by the spread of the Internet did not emerge.

This was caused by a combination of factors linked to limited re-regulation (which kept access prices high) and the demise of many dot coms (which dented user confidence in e-commerce).

Fourth, many overestimated the strength of the competitive positioning of new operators, their ability to win market share off incumbents and underestimated how aggressive the latter would be in seeking to ward off any competitors. This strategic problem was compounded by the commoditisation of network capacity. As a result there was very little to differentiate these firms other than price. The other problem was that the barriers to entry had been reduced with the result that competition has intensified and strategies duplicated. Furthermore these operators had built their business models on incurring substantial debt to build infrastructure which would be repaid as traffic rose. When this did not occur, their business models were undermined (Janisch, 2003).

As Rutherford *et al* (2004) noted the legacy of the extensive network build was that the main cities of northern Europe became the scene for very intense competition between operators which pre-empted an industry shakeout. The result was consolidation and a rapid decline in the level of private sector investment in network infrastructure. However, this has done little to reduce the vast levels of excess capacity of capacity represented by dark fibre. By 2005, only 3 per cent of European capacity is actually being used. In many cases, the cost of turning on this fibre may be more than the cost of actually burying the cable in the first place. Applying the technology to enable these networks (DWDM)[4] is very expensive. It was estimated by the beginning of 2002 that over $500 billion had been wasted on fibre cables that will never carry a single call or have been decommissioned (FT, 2001b). The problem is that when a firm goes bankrupt the capacity does not go away. The new owner will merely operate the infrastructure more cheaply as they are now free of sunk costs incurred by the network developer. This creates a contagion effect as the reduced prices push others towards financial difficulties.

The telecoms crisis and the impact on TENs

A by-product of the telecoms crisis was a growing concern as to the viability of a market-driven development of the information society. The financial difficulties of telecommunications operators have limited their ability and willingness to deploy the necessary services and technologies to support its development. As a consequence, at the state level, there was a desire – politically charged in some cases – to alleviate some of the debt accumulated by flagship operators so as not to endanger longer-term strategic objectives. Some states even questioned the deregulation process, claiming that it had generated unstable marketplaces at a time when greater stability was needed. The response of some governments was to make limited moves towards more active involvement in the sector. Most notable was the direct financial support offered by the French government to France Telecom who also called for a relaxing of the rules of state aid to the sector to enable broad strategic objectives to be met. There was even talk of renationalising some of these operators.

By the close of 2002, the Commission was openly considering ways that it could alleviate the difficulties within telecommunications without contravening its own rules, priorities and ensuring that the market-led route to the information society would not be endangered. Such actions included encouraging pre-competitive co-operation and enabling cost sharing in the deployment of key technologies. However in a review in December 2002, the Commission argued that many incumbents (the primary delivery method of the information society) came out of the telecoms crisis quite well or better than should have been expected. Many incumbents, given the demise of many new entrants and their control of the local loop (see below), sustained or even enhanced their dominance. For the Commission, challenging this dominant position was still the best way to pursue the development of the information society.

As a result, the Commission decided to pursue liberalisation with increased vigour. Ultimately, the Commission did not perceive that the longer-term objective of a market-driven information society was necessarily endangered by the telecoms crisis and concluded that more liberalisation was needed, not less. The Commission noted that if the sector is to come out of the period of slow growth, it needs to generate traffic and this is best achieved through more liberalisation. Thus the telecoms crisis led to the Commission reaffirming the liberal paradigm as the model for the development of the information society.

Whilst the liberal approach has produced a number of successes, the changes are incomplete as infrastructure issues remain in a number of key areas more especially in securing the spread of broadband which is central to the EU's strategy for the information economy. Solving these issues represents the key challenges to the liberal-led development paradigm as they require that the Commission not only re-enforces the liberalisation process (for the local loop), but also fosters co-operation (for advanced mobile services) and allows states to offer support (both financial and non-financial) to make the information society inclusive (the spread of broadband). Reconciling these competing demands is likely to dominate the telecommunications agenda for the foreseeable future.

The dilemma of the local loop

As there has been a sharp increase in capacity in national and pan-European backbones, so the focus of infrastructure development has shifted towards the local loop. This is the piece of infrastructure which connects the exchange to the end user. In competitive terms, the local loop is seen as an essential facility to which new entrants need access. Whilst the Commission has put broadband at the core of its policy strategy, the local loop represents an access bottleneck as there is a general absence of the necessary capable infrastructure to support the widespread deployment of advanced information society services/ applications. For competitive providers (as well as the Commission), it is the fact that the local loop remains under the control of the incumbent that has limited the roll out of broadband. The accusation is that incumbents have not rolled out broadband technologies fast enough and, through a reluctance to give up their last element of true monopoly control, are unwilling to let others undertake the task. In addition, where these technologies have been rolled out, only limited access has been offered to new entrants (Ure, 2003).

Removing the control by incumbents over this key part of Europe's information infrastructure has led many non-incumbents to call upon regulators to undertake local loop unbundling (LLU). LLU can take the following forms:

1. Full unbundling: selected local loops are taken over by a market entrant completely. The new operator is able to install its own equipment within the exchange to meet its customers' needs;
2. Line sharing: the same local loop is used by rival companies where the incumbent continues to provide basic voice services and the new entrants are able to use the local loop for more advanced services;
3. High speed bit stream access: the incumbent carries out the upgrade of the local loop and creates a link to the end user. The entrant is granted access to this facility.

However many incumbents have not only limited the ability of new entrants to build their own infrastructure by denying them access to their exchanges, they have also limited available capacity on their local networks (Gabelmann, 2001). The inability of new entrants to access exchanges to deploy their own infrastructures means that whilst many can and have built their own backbone, the sheer economies of developing an extensive system that connects every user, house or business to the network is beyond their financial capabilities. Thus new entrants need access to the local loop in fair and non-discriminatory terms.

The representatives of the network operators (ETNO) dismiss the need for LLU as part of the development of broadband within the EU. The incumbents argue that LLU is unnecessary due to the rapid pace of competition and technological change. In particular, the development of 3G (see below) removes the need for fixed access in the local loop. Over time, so the incumbents argue, as mobile links replace fixed links, there will be little need for LLU. Thus advocates for LLU base their analysis on an assumption that this is the only access technology. The development of new access technologies such as cable and fibre underline that there is no 'natural' way to enter the market. Evidence across Europe suggests that the deployment of broadband tends to depend on a multitude of approaches not a single method. Thus as incumbents have stalled for so long, the need for unbundling has been bypassed by the emergence of new technologies, etc. As Europe's unbundling was focused on old technologies, it could not provide the quality of service desired so new technologies need to be found.

On the whole, most EU states see LLU as a route towards securing the infrastructure to move to the information economy. Across the EU, LLU has depended upon (de Bijl and Peitz, 2005):

- The extent to which incumbents have upgraded their facilities to offer increased capacity on the local loop and have opened upon this facility to rivals;
- Progress in achieving consensus on co-location procedures and pricing;
- Availability of co-location space in those exchanges where multiple operators seek access;
- The pace at which regulators have sought to solve problems on the LLU;
- The rebalancing of prices to ensure that there is no price squeeze on new entrants.

Many incumbents have employed compression technology (such as ADSL)[5] which seeks to increase the capacity of existing infrastructure and means not having to build a whole new infrastructure (ETNO, 2000a, b). This was expected to broaden the appeal and development of new networks. The Commission also sought to push ADSL as a means of speeding the deployment of broadband. However incumbents have denied the roll out of ADSL by competitors through limiting the access to exchanges by new entrants thereby limiting the ability of its rivals to implement their own investment plans. This was compounded by incumbents pricing at a rate which was more attractive to their own customers than it was to wholesale customers.

ECTA,[6] as the representative of the new entrants, feels that despite the opportunities created by the new regulatory framework, the operators have been let down by the NRAs which have been ineffective (ECTA, 2003). This was often due to a lack of political credibility with central governments often failing to support the effective implementation of a competitive market or a direct challenge to the incumbent. Gradually ECTA has shifted emphasis from seeking LLU towards ensuring that wholesale access is fair. In part, this represents a failure by these operators to get effective movement in LLU by both the incumbent and the regulator. ECTA seeks a multi-technology route to broadband of which the local loop is just one method. It is evident to ECTA that the deployment of broadband is linked intimately to the effectiveness of the regulatory regime as delays in transposing directives into law play into the hands of the incumbent operator. ECTA sees this as important if the market-led route to the information economy is to be realised. These businesses have extensive investment plans for broadband but are being delayed by regulatory and commercial inertia.

Mobile network development and the European information economy

Mobile communications are central to the development of the European information society. In practice, this means migrating users (and applications) from current second generation (2G) mobile systems to advanced third generation (3G) technology (CEC, 2004c). In each state, the right to operate 3G services was sold to operators *via* various processes including auctions and beauty contests with many bidding large sums of money for these licences. Overall, across the EU-15, operators collectively paid national governments over €110 billion for 3G licences: though the Commission estimates the figure to be closer to €130 billion. This cost of the right to provide a service is compounded by infrastructure costs of €105 billion – a figure inflated by the conditions imposed by regulatory and licensing bodies (CEC, 2003).

However practical problems in the market for 3G services have limited its roll out (CEC, 2002), including:

- Delays in supply and demand as suppliers underestimated the time to get hardware ready and overestimated demand for these advanced services;
- Falling traffic expectations for traffic when the dot com bubble burst;
- Technical issues of speed and data from handsets in short available bandwidth;
- The lack of a harmonised stable regulatory framework, most notably towards the regulation of MVNOs which were allowed in some states but banned in others;
- Marked differences in licensing procedures between states which led to different market conditions between states;
- Uncertainties over the regulatory structure;
- Lack of financial resources;
- Environmental and health regulations notably over the positioning of masts;
- An inadequate number of licences and allocation of spectrum;
- Coverage obligations.

These problems were compounded by the high (and rising) debt incurred from attaining licences and created a situation where many of the operators who bid for the right to operate 3G networks could not afford to develop the necessary supporting infrastructure. ETNO partly blames this problem on a regulatory framework that limited the freedom of operators in terms of cost increasing and reducing measures such as coverage requirements and infrastructure sharing. For example, in some states where coverage requirements are not met, the operator will face sanctions.

One possible solution to this problem is to allow operators to share infrastructure. For operators, infrastructure sharing is a key issue in enabling the rapid introduction of 3G. It is estimated that through sharing, costs could be reduced by as much as 40 per cent, new technologies could be introduced sooner and environmental health concerns could be mitigated. This is especially pertinent given the need (under the terms of the licence) for extensive coverage. Network sharing has four basic degrees of co-operation:

- Shared use of suites (e.g. individual masts to sharing of grids);
- Shared use of base stations, antennas and radio centre controllers;

- The sharing of core network, including switching centres;
- The sharing of frequencies.

Whilst some feared network sharing represented an erosion of the liberal route to the creation of the information society (through potentially limiting competition and choice), the European Commission dismissed these concerns as unimportant especially in those areas with lower population densities.

Whilst conditions vary across Europe, generally infrastructure sharing is allowed if each operator maintains full operational control of their respective network. In the UK, sharing is limited to smaller cities. This reflects a need to both sustain competition and ensure that there is extensive coverage. Typical of the agreements to share infrastructure is that between MMO2 and T-Mobile where there was agreement to co-operate in both the UK and Germany. The combined savings will be in the region of €5 billion. In Germany, such agreements have saved operators an estimated €14 billion. By allowing such agreements, the European Commission created space for the development of 3G infrastructure. This demonstrates that the liberal approach normally taken by the Commission can be mitigated in meeting the long-term objectives of the European Commission. However these objectives have to be weighed against a desire for maximum choice for the consumer.

In addition to network sharing, there have been a number of other adaptations to the regulatory environment notably through changes to roll out obligations, licence duration and fees and other payments. These shifts in the licensing conditions were driven by a desire to meet the broad socio-political objectives of the public bodies to make market strategies less constrained. Roll out deadlines were extended when it became apparent that not only was equipment unavailable but also that the financial cost was, over the short term, prohibitive. However, the Commission has continued its focus on operators getting their market strategies right. The EU feels that its regulatory framework is sufficiently flexible to adapt to changing conditions but it has stated that there may need to be greater co-ordination between states. In addition, pressure is upon local bodies to improve the planning process for base station deployment. This is likely to require greater harmonisation and a speeding up of the procedures to develop base station sites. This has to be part of a broader harmonisation process where the market for mobile communications can be integrated across Europe. The longer-term impact is expected to be a consolidation within the industry.

Table 4.3 3G operators offering commercial services (September 2005)

State	Trial	Commercial-Trial	Commercial	Total
Belgium	2	1	0	3
Czech Republic	0	2	0	2
Denmark	0	0	1	1
Germany	0	0	4	4
Estonia	0	1	0	1
Greece	0	0	3	3
Spain	0	0	3	3
France	0	0	2	2
Eire	0	1	2	3
Italy	0	0	4	4
Cyprus	0	0	1	1
Latvia	2	0	0	2
Lituania	0	0	0	0
Luxembourg	0	1	2	3
Hungary	0	0	1	1
Malta	0	0	0	0
Netherlands	0	0	2	2
Austria	0	0	5	5
Poland	0	2	0	2
Portugal	0	0	3	3
Slovenia	0	0	1	1
Slovakia	0	0	0	0
Finland	1	0	2	3
Sweden	0	0	0	4
UK	0	2	3	5

Source: European Commission.

According to Carlberg (2001), the EU has been far too active in mobile markets and this activism has contributed to the sectors' problems through two main channels. First, the imposition of short-term deadlines pushing for the speedy creation of a 3G market resulted in businesses and other financial bodies over-committing themselves to a market without critical mass let alone maturity. Second, the desire of states to maintain control over the licensing and the development of the market undermined the EU's commitment to liberal markets. This may allow states to move towards more interventionist policies later. The state of roll out (as of September 2005) is highlighted in Table 4.3.

The challenge to net neutrality

The need by network operators to not only combat the commoditisation of network capacity but also to invest heavily in more advanced infrastructure has led to many – both in the US and Europe – to challenge the principle of net neutrality. This principle asserts that network service providers should not be allowed to dictate how the infrastructure is used with traffic treated in a non-discriminatory manner. The advantages of such an approach lie in the Schumpeterian notion of innovation as the dynamics provided by competition push advancement in the capabilities of these networks. However, as network operators are now private concerns, they are under an imperative to maximise the revenue from their core asset: their infrastructure. On this basis, operators have started to assert that uniformity of treatment is no longer warranted when the level of service offering begins to differ between users.

As it has evolved, so the demands placed upon the Internet have increased as more and more bandwidth hungry services (such as digital downloads, online gaming, etc.) emerge to fill up network capacity. To network operators, these services use up capacity to which these users were not entitled and, therefore represent in extreme cases a 'theft of service'. Initially any attempt to charge large users (such as Google, Yahoo Amazon and eBay) for the extra capacity used was stymied by regulatory authorities notably in the US who established 'network freedoms' which allowed users to access lawful content, run applications of their choice, connect legal devices to the network and for users to have choice between providers. The regulatory authority believed that net neutrality allowed a free market to work by allowing the best applications and services to flourish thereby stimulating innovation.

However network owners retain a desire to discriminate between different types of traffic. Network operators claim that no other network infrastructure operates in such a manner. Others (such as transport) enable providers to discriminate between users, especially in terms of the speed and quality of service. Operators feel that non-discrimination between different types of traffic does not answer the question as to how to fund the supporting infrastructure. The emergence of certain types of high impact traffic places more demands upon the network especially within the local loop which needs to be upgraded if service quality is going to be maintained for all users. Operators argue that in order to give the service these enterprises need, they need to guarantee a higher level service through securing the necessary bandwidth and therefore need to charge more. Thus in order to fund an infrastructure with high service levels that is free from congestion, operators need to discriminate between different types of users.

This argument has now spread to Europe where many incumbent operators are seeking the power to bring these rules into place (FT, 2006). Operators are planning to invest over €80 billion in advanced networks in the four years to 2010 and believe that profitable Internet companies should be expected to pay a higher amount to help pay for the networks that they are filling up with their content. Deutsche Telekom and Telecom Italia are pushing for an end to net neutrality within their respective domains. Other incumbents are more ambivalent with BT saying it has no complaints against Internet companies (though it is open to persuasion on the issue) and France Telecom dismissing the issue of net neutrality as an American debate.

The Internet companies respond that they already pay 'peering rights' to the network owners. These peering rights are based on the bandwidth amount they use. Clearly operators can get around the revenue problem by raising these peering right charges and consumer access fees. In combination, these could fill funding gaps without stifling innovation. Some fear that these moves by incumbents are part of a defensive strategy to ensure that their traffic is given priority over those of their emerging rivals. Many incumbents have plans to move into TV over the Internet. Limiting access to others could give these operators preferential access and dominate this market segment. In this context, the move to end net neutrality whilst being a symptom of a shift towards a more liberal regime could also be about seeking to retain incumbent position through limiting the ability of rivals to establish a credible presence within this market segment.

Public policy measures towards the development of trans-European telecommunication networks

Policy actions at the supranational level are set within the broad generic market-led framework outlined above. Given the limited financial resources at its disposal, regulation is perhaps the most effective means that the Commission has to influence the development of the European information society. More direct action within information markets has focused upon:

- Sectoral stimulation: based on supporting an indigenous market for electronic services by developing measures on both the demand and supply side;
- Exploiting the benefits of the information society by ensuring that the technologies and services related to the development of the

information society diffuse throughout the European socio-economic body.

EU activity in telecommunications, and more latterly the information society, has generated a plethora of initiatives crossing many activities of the Union from transport through to cohesion. Since 2000 (and up to 2005) these initiatives have been consolidated under a single initiative: eEurope (see below). This has set a goal of creating a virtuous cycle of market-driven forces for information society products/services in the EU. This market-driven strategy has been complemented by the EU's ongoing research efforts and actions in developing what can be broadly defined as 'socially relevant' content and services. On the supply side, the policy sought to remove all remaining obstacles to the upgrading and deployment of infrastructure and to ensure that financial support from public bodies addresses clear cases of market failure. On the demand side, policy sought to remove regulatory obstacles to demand for new services, encourage users to exploit the technology, facilities adoption and pool demand where possible to create an effective demand for the necessary services.

Often these policies have only an indirect influence on the development of advanced infrastructure through either filling up available capacity or by creating the demand for networks to stimulate further investment. The initial efforts in developing infrastructure *via* EU policies (under the TEN-Telecom programme) have moved on from trials, interconnection and development towards more service-based deployment and away from developing physical infrastructure *per se*. This shift in policy reflects the substantial investment in infrastructure that has taken place over the last two decades. The policy ties in with research initiatives to ensure that trials can be rolled out based on a desire to minimise the risk of developing and deploying those services/applications that are pivotal to the socioeconomic development of the information economy. This underlies the core objective of policy by ensuring that the capacity of infrastructure is partly filled with content that is socially progressive and commercially neutral. It also reflects a desire to ensure that traffic is generated for major pan-European backbones by supporting transnational services. Public bodies are keen that the information society must not only entertain but also embed itself within socioeconomic and political structures. This has led the EU to becoming involved in content development. This is one of the more controversial areas as the EU is getting involved not only in facilitating network development but also in what is delivered over them.

A central aim of the policy is to avoid failures in the industrial age reasserting themselves in the information economy. Thus there is the desire to ensure employability, to develop new skills and to address any other social gaps that could emerge from the evolving information society. However Melody (1996) criticises the fact that information society policies have an uncritical acceptance of the benefits of the development of the information economy. Policy also asserts a belief that information poverty is a bad thing and that all can adapt to its challenges. Thus there is a desire for the development of the information society to be demand led (as opposed to being supply forced) if social and economic efficiency is to be maximised.

The aforementioned involvement in content markets has been driven by a desire to meet the broad socioeconomic objectives of the EU, notably in terms of culture, society and e-government. In other areas, there are efforts to increase its relevance through skills and regulatory factors to generate familiarity right across the socioeconomic spectrum. These themes have been continued under the new programme – i2010. In terms of infrastructure, this stresses the policy priority of creating a 'single European information space' based on familiar themes of access to broadband, rich content, interoperability and security. Once again, this puts broadband at the core of the strategy and as the building block for what follows. This is a market-led initiative in which the EU seeks to confine itself to what it sees as market failure. Thus the aim is to cover areas of socioeconomic importance that markets will not develop due to immaturity or outright failure and facilitate action in others by creating a common regulatory playing field. This latter point stresses broad international interaction.

However, despite the claims of an adherence to a liberal framework for the development of the information economy there is – within EU policy – a relatively strong interventionist and deterministic slant, most notably in terms of its involvement within content markets. Furthermore, many states within the EU have to effectively challenge incumbents who still control key parts of the national infrastructure (see above). The EU promoted the information society and justified its role in the development of the information society as a means of securing transnational dimensions in its development. However with increased interaction of economic actors across borders, there are questions over the degree to which such support is needed given the intensity of these interactions. When transnationality is a normal condition of commercial interaction, this reason for policy action may no longer be justified.

National broadband strategies

Spreading access to broadband technologies is central to the EU's strategy for the information society. It was feared that one of the immediate casualties of the telecoms crisis was that many operators – both new and incumbent – would scale back their plans for the widespread deployment of broadband networks. The Commission's response to this concern was to pursue liberalisation of broadband services more aggressively in the hope that this would stimulate the necessary investment (CEC, 2004a). To some degree, the Commission has been vindicated in this approach as, despite the telecoms crisis, broadband take up within Europe has maintained a steady progress. By the end of January 2006, there was estimated to be around 59 million broadband subscribers (a penetration rate of 13 per cent for the EU-15); an increase of over 46 million in three years (CEC, 2006). Whilst all EU states lag behind South Korea in broadband take up, in five EU states (Netherlands, Denmark, Finland, Belgium and Sweden) the penetration rate exceeds that for both the US and Japan (see Table 4.4). It is no accident that the roll-out of broadband is strongest in those states with the highest intensity of competition (CEC, 2005).

To offer focus and speed the roll out of broadband, the Commission has requested that member states establish their own deployment strategies. These national broadband strategies involve national governments establishing targets to stimulate a market for broadband services. These are often implemented by local bodies who are more in tune with the needs of local economy. To support the roll out of these technologies in these locations, there is a mix of both indirect (*via* re-regulation) and direct support which can include:

- Public funding: this could be in the form of loans or grants or *via* fiscal incentives to users;
- State aid: assistance to speed roll out of broadband with EU law. In 2005 and 2006, the Commission approved a number of state aid schemes to aid roll out of broadband in Spain, Ireland, Austria and the UK;
- EU public funding *via* devices such as the structural funds: the Commission has offered €210 million to finance broadband development in Greece, for example;
- Demand aggregation and procurement: state efforts to coalesce demand and aid purchase of core parts of the network;
- Fostering markets: this could focus on key areas to aid content in a number of public service.

Table 4.4 Comparative broadband deployment

Country	Broadband Population Penetration Rate (%) (June/July 2005)	Annual Increase in Penetration Rate (%)
South Korea	25.5	1.26
Netherlands	22.4	7.8
Denmark	22	4.9
Iceland	21.7	6.41
Switzerland	20.3	5.86
Canada	19.2	2.65
Finland	18.7	7.74
Norway	18.2	6.89
Belgium	17.4	3.5
Sweden	17.1	4.49
Japan	16.4	3.73
US	14.5	3.48
EU-15	12	4.4
Australia	10.9	5.68
EU	10.6	4.1
EU-10	3.1	1.8

Source: European Commission.

In practice, these strategies require policy makers to engage with the incumbent to stimulate national broadband roll out and take up. This interface with the incumbent reflects the fact that these firms still dominate the provision of broadband within Europe. As of October 2005, the incumbent has (on average) 50 per cent of the retail market for fixed broadband (see Table 4.1). Whilst this has fallen 6 per cent over the year, the remainder of the market does tend to be highly splintered between a large number of rival suppliers. Only nine of the EU-25 are in a position where new entrants collectively have a larger market share than the incumbent. This power is highlighted by the control that incumbents have over DSL (the main access technology for broadband) where these businesses have on average 60 per cent of market share. Indeed over half of the EU-25 incumbents have over 80 per cent of the market for this technology (CEC, 2005).

The key issue is whether this dominance matters in terms of broadband deployment. The EU has put the rapid roll out of broadband at the top of its policy priorities, driven by a desire to bridge an emerging digital divide across territories. Whilst the broadband market has matured, such maturity has been focused on the more developed regions of the EU with access to remote and rural regions remaining limited due to high costs and low population density. These remote and rural areas are also characterised by limited competition and the continuation of the incumbents as either the monopoly or dominant supplier. Implicitly, the active measures deployed by national and supranational bodies offer support to incumbents who dominate supply in these under-served areas. In placing the rapidity of roll out (rather than waiting for competition to mature) as its policy priority, local, national and supranational bodies have recognised that the incumbent is the most practical deliverer of the information society services and applications to remote and rural regions. New entrants are focused on high population areas and it is only the incumbent that has the physical presence in these locations. Whilst set within the context of a market-driven strategy, such activism (even though it may only be temporary) arguably represents a retreat away from the liberal policy. These measures stress a continuing interface between government industrial policy and the incumbent based upon the latter to achieve specified policy aims. This reliance opens up the possibility of more aggressive non-market strategies by incumbents where state assistance occurs alongside regulatory actions to aid its position in more competitive segments. Such links create effective political clout for incumbents and can imbue key aspects of their non-market strategies in trading off strategies for socioeconomic cohesion with a more regulated and elongated liberalisation process.

The nature of the government interface with the incumbent to create this socially inclusive vision is an important benchmark of the transformation of government policy. Such debates focus on the extent to which policy needs incumbents to deliver the information society. Thus governments face a choice of offering direct subsidies or allowing the incumbent to generate extra revenue in other segments on the basis that these monies are channelled into target areas and services. Alternatively, the state could actively encourage entry into these markets by emergent non-incumbents. Thus policy has to address whether to meet its priorities through incumbents or preferential entry to new entrants. The policy response will depend on the ideological slant of the government. However, there are concerns about states replacing markets and distorting markets through their actions. However, if they can separate wholesale and retail provision, then competition can be facilitated. Although given that incumbents have an extensive network, they appear to be the most natural conduits.

Incumbents take a very proactive role in socioeconomic development and work with governments to secure widespread access to broadband. This reflects both active market and non-market strategies. Having the most extensive networks, these are the operators most able to reach – through fixed line telephony – the most remote locations within national economies. In part, this reflects the universal service provision within the regulatory framework, though many are seeking to renegotiate better terms for the commercial obligation of universal service in an environment where mobile telephony is widely accessible and where competition is maturing. Often these social aspects will be linked to broader reforms. Thus for incumbents, digital inclusion is linked intimately into the regulatory regime where social commitments have to be set within a trade-off with regards to the asymmetric requirements that exist upon the operator. It also reflects the fact that the social aspect of telecommunications is not a priority for non-incumbents as these will merely seek to cream skim activities of the incumbent. However as markets mature, their presence in these areas and their ability to generate traffic could work to the benefit of incumbents. Thus for many states, there is little choice but to seek to realise the information society through the power of incumbents. This will give the development of the information society over the short term a more strategic perspective. Free competition has limited the ability to create a market-led information society and states will rely upon the strategic pattern of development. Most incumbents collaborate actively with the public sector and are making it an increasingly key focus of their strategy.

Conclusion: towards a new framework for the emerging information society

The last two decades have witnessed a pronounced shift in the network development paradigm for telecommunications infrastructures within the EU. The progressive liberalisation of the sector combined with the privatisation of the incumbent operator has highlighted the shift as services become more imbued with market-driven forces. In terms of liberalisation, the full framework has now by and large been fully accepted if not fully implemented across the EU. Currently of the EU-25 incumbents, seven have been fully privatised, only two (Cyprus and Luxembourg) are still fully in state ownership whilst a further four have state ownership of 50 per cent or more. The remaining 12 are now in minority state ownership. However, in some cases there is a residual golden share which can offer the opportunity for the state to veto the strategic actions of the incumbent (see Table 4.1).

Progress towards a liberal programme of network development has been gradual and is symptomatic of a shift towards more passive industrial policies. This gradualism has been formed by the uneven implementation of the deregulation measures by states as well as the gradualist approach taken by the European Commission. Within the liberal paradigm, investment in infrastructure will occur where traffic is generated. This means a focus on network development in urban regions with a risk that rural and less well off urban regions could be excluded from the development of the information economy.

Prior to the telecoms crash, there was widespread confidence that markets could allow the widespread development of the information society through a pluralist network of networks. However as traffic levels did not rise as fast as anticipated, as non-incumbents went bankrupt and as the market became more consolidated, new interfaces started to emerge between incumbents and the objectives of government policy. This consolidation has facilitated an increase in the power of incumbents, especially with regards to fixed line communications, so member states have come to realise that they can either wait for the market to mature to allow the pluralist development of the information society or they can utilise the power of incumbents to deliver the information society to all parts of the socioeconomic structure. This latter policy option has become more evident across the EU where to some degree, exemptions from competition law or regulation have been allowed in return for promoting socially inclusive infrastructure developments.

The measures have also been evident in other areas of policy where the Commission as well as member states have followed a more strategic line. The Commission has allowed closer collaboration in infrastructure development if it is consistent with the broad objective of economic and social progress. Efforts within mobile telephony are allowing collaboration between operators in the development of infrastructure to ensure that these market entry concerns do not delay the roll out of these technologies. Thus the EU has shown the need to be flexible over the form and pattern of network development in pushing the economic priority of stimulating the development of the information society. Thus in some segments of the telecommunications industry, there has been a limited move towards greater activism through policy measures and monopoly *via* consolidation. This move away from the pure liberal paradigm may be temporary but it reflects the prevailing uncertainty as to whether market forces would be able to address the priorities and needs of the information society.

References

Berben, C. and Clements, B. (1995) 'The European framework for competition in telecommunications: the benefits for peripheral countries', *Telecommunications Policy*, 19(4): 273–83.

Berhin, D., Frédéric Godar, F., Jollès, M. and Nihoul, P. (2005) 'Sector-specific regulation in European electronic communications – meant to disappear?' *Info: the Journal of Policy, Regulation and Strategy for Telecommunications, Information and Media*, 7(1): 4–20.

Carlberg, R. (2001) 'The persistence of the dirigiste model: wireless spectrum allocation in Europe a la francaise', *Indiana Law Journal*, 54(1): 129–64.

Commission of the European Communities (CEC) (1998) *The Competitiveness of European Enterprises in the face of globalisation – How can it be encouraged*, COM(98) 718.

Commission of the European Communities (CEC) (2002) *Towards the Full Roll-Out of Third Generation Mobile Communications*, COM(2002) 301 final, Brussels.

Commission of the European Communities (CEC) (2003) *Electronic Communications: the Road to the Knowledge Economy*, COM(2003) 65 final, Brussels.

Commission of the European Communities (CEC) (2004a) *Connecting Europe at High Speed: National Broadband Strategies*, SEC(2004) 599, Brussels.

Commission of the European Communities (CEC) (2004b) *Connecting Europe at high speed: recent developments in the sector of electronic communications*, COM (2004) 61 final, Brussels.

Commission of the European Communities (CEC) (2004c) *Mobile Broadband Services*, COM(2004) 447 final, Brussels.

Commission of the European Communities (CEC) (2005) *Broadband access in the EU: situation at 1 January 2005*, COM(2005) 12 final, Brussels.

Commission of the European Communities (CEC) (2006) *European Electronic Communications Regulation and Markets 2005*, COM(2006) 68 final, Brussels.

Daßler, T. and Parker, D. (2004) 'Harmony or disharmony in the regulation and the promotion of competition in EU telecommunications? A survey of the regulatory offices', *Utilities Policy*, 12(1): 9–28.

de Bijl, P. and Peitz, M. (2005) 'Local Loop Unbundling in Europe: Experience, Prospects and Policy Challenges', *Communications & Strategies*, Issue 57, 1st quarter: 33–59.

The Economist (2002) *The Great telecoms Crash*, 18 July 2002, www.economist.com.

European Competitive Telecommunications Association (ECTA) (2003) *ECTA Sustainable Broadband position*, www.ectaportal.com.

European Telecommunications Network Operators (ETNO) (2000a) *Reflection Document on the 'Draft EU Regulation on Unbundled Access to the Local Loop COM (2000) 394'*.

European Telecommunications Network Operators (ETNO (2000b) *Reflection Document on 'The Commission Recommendation on Local Loop Unbundling (Working Document 09.02.2000)'* www.etno.be.

Financial Times (2001a) *Glorious hopes on a trillion-dollar scrapheap: The Telecoms Crash – Part I*, 5 Sept 2001, p.12.

Financial Times (2001b) *How the world caught net fever: The Telecoms Crash – Part II*, 7 Sept 2001, p.12.

Financial Times (2001c) *The tangled legacy of a derailed revolution: The Telecoms Crash – Part III*, 8 Sept 2001, p.13.

Financial Times (2006) *A two speed internet? Why network operators are flexing their muscles*, 29 March 2006, p.15.

Fransmann, M. (2002) *Telecoms in the Internet Age: From Boom to Bust to ... ?* Oxford: Oxford University Press.

Gabelmann, A. (2001) Regulating European telecommunications markets: unbundled access to the local loop outside urban areas', *Telecommunications Policy*, 25(10–11): 729–41.

Garnham, N. (1997) 'Europe and the Global Information Society: the history of a troubled relationship', *Telematics and Informatics*, 14(4): 323–7.

Janisch, H. (2003) 'Telecommunications in turmoil' *Conference on 'Competition, Pricing and Natural Monopoly: Taking Stock of Network Industries'*, Centre for Innovation Law and Policy, University of Toronto, 23 May 2003.

Jordana, J. (ed.) (2002) *Governing Telecommunications and the New Information Society in Europe*, Cheltenham: Edward Elgar.

Larouche, P. (2000) *Competition Law and Regulation in European Telecommunications*, Oxford: Oxford University Press.

Liikanen, E. (2002) 'The eEurope Broadband Strategy', *The European Telecommunications Network Operators' Association (ETNO) Conference 'Making Broadband Happen in Europe'*, Brussels, 3 December 2002.

Mansell, R. (1993) *The New Telecommunications: A Political Economy of Network Evolution*, Thousand Oaks: Sage.

Melody, W.H. (1996) 'Toward a framework for designing information society policies', *Telecommunications Policy*, 20(4): 243–59.

Natalicchi, G. (2001) *Wiring Europe: Reshaping the European Telecommunications Regime*, Lanham: Rowman & Littlefield Publishers.

Noam, E. (1992) *Telecommunications in Europe*, Oxford: Oxford University Press.

Noam, E. and De Sola Pool, I. (eds) (1980) *Technologies with Boundaries: On Telecommunications in a Global Age*, Boston, MA: Harvard University Press.

Preston, P. (1995) 'Competition in the telecommunications infrastructure: implications for the peripheral regions and small countries in Europe', *Telecommunications Policy*, 19(4): 253–71.

Rutherford, J., Gillespie, A. and Richardson, R. (2004) 'The territoriality of pan-European telecommunications backbone networks', *Journal of Urban Technology*, 11(3): 1–34.

Ure, J. (2003) 'Competition in the local loop: unbundling or unbungling?' *Info: the Journal of Policy, Regulation and Strategy for Telecommunications, Information and Media*, 5(5): 38–47.

Waverman, L. and Sirel, E. (1997) 'European telecommunications markets on the verge of full liberalization', *The Journal of Economic Perspectives*, 11(4): 113–26.

5
Energy TENs

Energy is widely regarded as a sector of broad strategic importance, a characteristic that helps define it as one of Strange's secondary power sectors and keeps it, despite shifting ownership and liberalisation, as a focus of public policy concern. Within the European context, energy has long been considered as a vehicle for European integration. It figured prominently in two out of the three founding treaties of the European Communities – the 1951 Treaty of Paris which set up the European Coal and Steel Community (ECSC) and the 1957 treaty establishing the European Atomic Energy Commission (Euroatom). The neo-functionalist aspirations of the founding fathers envisaged that the co-operation in the key coal and steel sectors would stimulate closer integration in other sectors, setting off a snowball effect that would stimulate integration in related areas across an increasing range of activities.

The three driving themes of EU energy policy have traditionally been security of supply, the environment and competition. Supply security (that is, ensuring reliable and good quality supplies for European markets) has always been at the forefront of policy concerns[1] when there has been political instability in the producing areas – which for oil and gas have tended to be far distant from the main consuming areas in Europe. Given the absolute reliance of the modern economy on both the quantity and increasingly the non-interruptibility of energy supplies, there has always been a strong element of control in energy policy.

The growth of liberalism represents a shift in thinking about supply security. Increasingly, the market has been regarded as a better mechanism to resolve key energy challenges (although, as seen below, in some cases this has often been in theory rather in practice). Market access, both within and between markets, has been the watchword of this more liberal approach. This market access has increased the energy interdependence between member states which, in turn, has led to a rethinking of supply security. That is, if there is a supply crisis in part of the European market, the interconnectedness of key networks, the major feature of energy TENs, will bolster supply security as shortages can be offset by the dispatch of supplies from elsewhere in the Community.[2]

Thus, the monopoly control of the state over electricity and gas has been undermined by the shift in thinking (see below) whereas the European Commission has been seeking a co-ordinating role for itself in which it helps create a more market-driven environment. TENs, which has a more liberal aspect in energy than in the transport sector, for example, where the role of the state remains crucial in many sectors for financial reasons, is seen as an instrument to deliver both security of supply and more competitive markets. Undoubtedly, new infrastructure is needed to bring gas in particular into the Community. However, it remains to be seen how states will react if anything but the most temporary supply crisis occurs.

This chapter develops the idea of energy, notably electricity and gas, as necessary underpinnings for modern, interdependent economies. In particular, it highlights how the construction of transnational energy networks cannot be separated from broad aspects of energy policy, especially the intertwined nature of TENs and energy market liberalisation, and increasingly security of supply. In short, proper energy liberalisation requires the removal of all obstacles to cross-border trade, not only legal obstacles but also physical obstacles in the form of bottlenecks, missing links and the removal of congestion. Conversely, without the incentive of legally open markets, the impetus to build new connections is reduced. Thus, the chapter begins with a section on the general characteristics of gas and electricity networks. It then discusses the evolution of energy TENs to date before examining the liberalisation focus in Europe with an emphasis on the extent to which liberalisation has succeeded, particularly in relation to its contribution to TENs. The final section pulls together the earlier themes and sets them within the context of the strategic framework set out in Chapter 1.

General characteristics of energy networks

The integrity of energy networks is crucial to the smooth functioning of modern economies. Gas and electricity grids are the means by which energy is transported from the point of production to the point of consumption. Power generation and gas production need the appropriate wires and pipelines to transport the energy to the end-user. The energy transported over the network is a fundamental economic resource which underpins all forms of social and economic activity: energy supplies heat, light and mobility and enables industry to transform raw materials into finished products.

In addition, the energy utilities are also of major importance in their own right in terms of their total sales, assets and employment. In 2001, for example, the value of final energy demand was equivalent to 4.3 per cent of gross value-added in the EU-15 (CEC, 2004b). However, unlike most other economic sectors, the internationalisation of these utilities is relatively underdeveloped. In part, this is because, for many years, network infrastructures were shaped by local, regional or national monopolies and were contained within these fragmented frameworks. Although trade between networks did take place, it was marginal. Electricity, for example, was directed towards helping individual networks attain constant balance between supply and demand. Arguments for greater interconnection on the grounds of security of supply or to intensify competition were not popular in a control-driven environment. Incumbents, in particular, emphasised the naturally monopolistic nature of their industry to maintain their pre-eminent position. It was only when the strong belief in the inevitability of natural monopolies in energy networks broke down that opening these markets to new entrants and larger amounts of trade became not only possible but desirable (see below), at least in the eyes of some governments.

Liberalisation of Europe's gas and electricity sectors is a key factor in the internationalisation process of which TENs represents one of the first tentative steps. TENs is also an important part of the regime change represented by the shift away from the process of control, facilitating the development of demand-driven networks that are based on alternative forms of development to state/public sector provision, including the involvement of private and non-national parties in their development.

Despite longstanding cross-border transmission, exchanges between national energy systems were limited and fell far short of underpinning a genuine pan-European energy market. Both networks currently require legally unbundled networks to which operators can gain access to efficient transportation on a non-discriminatory and transparent basis. This represents a major change around in thinking.

Both electricity and gas networks are regarded as key infrastructures in which a lot of political capital has been invested, both in terms of the imposition of public service obligations in support of social and regional cohesion objectives and more broadly in terms of ensuring there are sufficient energy supplies to sustain economic activity. Liberalisation has facilitated the emerging of energy trading and markets but, as discussed below, significant obstacles remain before full market integration is achieved. Given the importance attached to energy, it is perhaps unsurprising that adoption and acceptance of the liberalisation view has taken place at a variety of rates within Europe.

Important differences exist between the transportation systems for different energy sectors. Electricity transportation is entirely line-bound. Gas transportation is predominantly line-bound (pipelines) with some gas being transported by LNG carrier, a mode of transport that requires gas liquefaction and gasification infrastructure at departure and arrival points respectively. Waterborne gas transportation has long been uncompetitive with pipelines but technological developments have started to redress the balance. Oil[3] can be transported relatively easily by road, rail, ship or pipeline. The choice of oil transportation mode and the route are determined by economics, geography and geopolitics.

Differences in gas and electricity networks stem from their basic characteristics. Gas (or at least natural gas) can only be produced where it is found whereas electricity can be generated anywhere. Consequently, gas transportation tends to take place over much longer distances (often thousands of kilometres) compared to electricity which incurs losses over distance, making it more rational to transport the primary fuel rather than electricity over distances of more than a few hundred kilometres. Thus, proportionally more of the gas consumed in Europe crosses borders than electricity: over 60 per cent of gas consumed in Europe is transported across borders compared to about 9 per cent of electricity. The emphasis therefore in fostering trans-European gas networks has been on securing and diversifying gas supplies from outside the European Union whereas the impetus behind electricity networks, given that most member states are self-sufficient in power, has been on fomenting competition and on balancing short-term supply and demand.

Moreover, gas is a primary energy[4] source which is in competition with other fuels in many applications whereas electricity is a secondary energy source and faces less competition in most usages. Electricity is a universal service, placing obligations on the electricity supply industry which do not apply to the gas industry which is not a universal service. As gas can be stored, gas storage facilities are an important part of the industry's infrastructure, enabling markets to withstand short-term interruptions to supply. However, electricity cannot be stored, necessitating constant balancing of supply and demand. The scale and cost of investment in both sectors is high but electricity tends to encounter greater opposition to the construction of its basic generation and transmission infrastructure and thus longer delays to the authorisation of new projects. Long-term contracts exist in both sectors but have been particularly important for the gas industry in getting large-scale investment projects off the ground.

A fundamental rethink of the nature of the line-bound energy industries has occurred since the 1980s which has resulted in an elemental restructuring of the industry – in both a vertical and horizontal fashion – with major implications for all stages of the value chain. Previously, electricity and gas supply were considered natural monopolies. Disaggregation of the functions of the utilities changed all this. The main stages in the production of both gas and electricity were all affected by this process of change:

- Generation: that is, production of the energy itself;
- Transmission: that is, long distance transport of gas and electricity through high pressure pipelines and high voltage grids respectively;
- Distribution: that is, the transportation of gas and electricity from the transmission grid to the end-user through low pressure pipelines and low voltage grids respectively;
- Supply: a wholesale and retail trading function which includes sales, metering, invoicing and collection.

According to the new thinking, generation and supply were considered amenable to competition whereas the line-bound aspects of the industry, particularly transmission which lies at the heart of the TENs initiative, were still considered natural monopolies as it was commercially non-viable to build duplicate networks through which to compete. However, devices such as third party access (TPA) offered the potential to open up the most monopolistic part of the respective industries. Provided producers and their customers could gain access to the networks on fair and equal terms, even if the networks themselves remained under monopoly control, competition became possible in these sectors.

It is access to networks and the configuration of networks that holds the key to competition throughout the European energy network industries. In order to achieve this, consideration of the network quality characteristics outlined in generic terms in Chapters 1 and 2 need to be taken into account within the specific context of the energy network industries. For example, interoperability and interconnection between separate networks pose technical, albeit not insurmountable, challenges for electricity supply. Flexibility to respond to changing needs requires adaptation to new demand patterns. In the gas sector, for example, the depletion of the UK North Sea gasfields necessitates new connections[5] and sources of supply if the UK is to continue to meet its domestic gas demands. The 2003 blackouts in Italy demonstrated the need for the reliability and continuity of the networks and regional and economic imperatives, among other factors are determining the geographical reach of networks. The way in which network congestion is handled and the accessibility of gas and electricity networks can either dampen or enhance the efficacy of network competition. Missing links in the network chain affect the network. Located in the middle of the European Union but isolated by its alpine geography, Austria, for example, poses some problems in this respect.

Other considerations also apply to the development of trans-European energy networks. For example, integrated power networks enable utilities to take advantage of complementarities in generation. The most integrated part of the European network is in the Nordic region where hydro generation is cheap and plentiful at certain times of the year but where thermal generation can provide the base load when hydro power is not so plentiful. Moreover as the networks develop, the EU is increasingly reaching out beyond its borders to integrate its markets with those of its neighbours.

Evolution of policy towards energy TENs

The formation of integrated and seamless pan-European electricity and gas networks derives from two of the EU's main energy policy objectives – the creation of competitive and efficient single energy markets and security of supply. Both objectives have played their part in promoting regime change. The part played by energy sector liberalisation in the shift away from control to a more passive policy framework is clear. The utilisation of TENs to improve energy security represents a new approach to this increasingly important issue and again calls upon the market, if not to resolve security issues entirely, at least to make a major contribution to achievement of this policy objective.

In order to achieve the liberalised markets and the resulting intensification of competition, lower prices and greater efficiency alluded to in the previous section, the interconnection of hitherto largely nationally configured networks needs to be intensified. However, an integrated network will only be constructed if access to the networks is non-discriminatory and easy. In other words, the more level the playing field, the greater is the incentive to connect all areas on it. Thus, the energy TENs initiative faces a 'chicken and egg' dilemma: an integrated network is essential for the creation of a single energy market or else the EU could be left with a series of separate national networks (some of which are too small or too isolated for a competitive network) – some of which may be competitive internally but others of which will not. However, the incentive to develop the integrated network ideally requires the liberal framework to be in place. As such, the two initiatives are inching forward bit by bit with interconnection stronger in some regions of the EU than in others. The problem facing the Commission is that some states are more committed than others to simultaneous achievement of liberalisation and interdependence. In states where the incumbent retains much of its influence, the incumbent is frequently able to benefit from liberalisation and interconnection without any undermining of its own domestic position, thereby undermining the liberal aspects of the European policy.

The European Union's preoccupation with energy infrastructure stems back to the early 1960s when pipelines, which were relatively new to Europe, were regarded as a transport mode which had the potential to divert energy goods from more traditional transport means such as road, rail and sea. However, in practice, the Community's limited infrastructure initiatives during its first two decades had even less impact on energy than on more traditional transport modes.

The first infrastructure initiative relating specifically to energy occurred in 1976 and was prompted by security of supply concerns arising out of the 1973 oil crisis. The 1976 Council Regulation (OJL, 1976) required member states to inform the Commission of all investment projects[6] of Community interest in the petroleum, natural gas and electricity sectors. The purpose of the Regulation was to provide the Commission with a detailed overview of planned developments in capacities and equipment within the Community's energy sector. The notification of investment projects reflected similar provisions in the Euratom and ECSC Treaties to provide data for evaluating, and, if appropriate, influencing these investments. This was a limited attempt by the EU to challenge the extensive control at work in these sectors and to attain for itself a co-ordinating and consultative role. The big assault on control had to wait until the first attempts to liberalise energy markets at European level.

Since the relaunch of infrastructure policy in the late 1980s, energy has been central to TENs policy. The landmark 1990 Action Programme (CEC, 1990) contained the Commission's early thinking on energy networks. Among other things, the Action Programme identified priority action areas (see Table 5.1) – priorities which were later refined and developed in the energy guidelines and the Christophersen Report and included greater interconnectivity for more isolated powers networks

Table 5.1 Energy priorities identified in 1990 TENs Action Programme

Electricity interconnections:
New links:
Ireland–United Kingdom (1995)
Greece–Italy (1993)
Reinforcing interconnections (1991–95) between:
France and Spain–Portugal
France and Germany–Belgium–Italy–Switzerland
Italy and Austria–Switzerland
Germany and Netherlands–Denmark–Austria
Reinforcing internal links (1991–95) in:
Germany, France, Belgium and the Netherlands
Gas interconnections:
Introduction of natural gas in:
Greece (1993) and Portugal (1994)
Interconnections:
Ireland–United Kingdom (1993)
France–Spain–Portugal
Belgium–Germany
Germany–Denmark
Denmark–Norway ('Scanpipe')
United Kingdom–Continent
Sardinia–Corsica and mainland Italy
Trans-Mediterranean links:
Reinforcement of Transmed pipeline (Italy–Tunisia, 1992)
Interconnection of Spain and Morocco (1995)
Three projects in Germany to link up the new länder (1991–93)
New ways to bring in Norwegian gas:
Midal projects between Emden and Ludwigshafen and 'Zeepipe' (Belgium)

such as Ireland and Greece, stronger interconnections between member states and the strengthening of links within certain member states that would enhance the performance of the wider network. In relation to gas, the emphasis was on introducing gas into countries like Greece and Portugal, enhancing links within Europe and, significantly for security of supply concerns, reinforcing links between member states and third country producers, particularly from North Africa and Norway.

In early 1992, the Commission published the draft Regulation on Declarations of European Interest (DEI) (CEC, 1992a) in relation to the establishment of trans-European gas and electricity networks. This proposal mirrored DEI proposals for transport and telecommunications and shared their aims of trying to speed up the implementation of infrastructure projects by creating political commitment by the public authorities to specific projects.

The key document, the Communication from the Commission *Electricity and Gas Transmission Infrastructures in the Community* (CEC, 1992b) was published in March 1992. Prior to this, details of the trans-European energy networks had been sketchy. The Infrastructure Communication took stock of the existing configuration of both networks and examined how national infrastructures could be transformed into networks with a European dimension. In short, the document represented the first stage in the development of guidelines regarding the objectives, priorities and identification of projects of common interest envisaged by Article 129C of the Treaty on European Union.

These early stages of the energy TENs initiative took place within the context of the regime of control. Energy liberalisation, the parallel policy to the development of energy TENs, took a long time to gain acceptance and, indeed, agreement on the first stage of electricity liberalisation within Europe was only reached in 1996. In effect, early TENs was geared towards enabling national incumbents to establish interconnection. There was some activism and liberalism in a minority of states but, at this stage, energy TENs were to be delivered *via* prevailing political and economic structures.

For the first time, the 1992 Communication set out the specific energy advantages, as well as the more integrative benefits, for greater gas and electricity interconnection. It makes the point that existing electricity and gas grids had been developed to achieve national self-sufficiency and, although some international co-operation exists, the emphasis on grid development had been mainly national. This preoccupation made it difficult, first to reap the full benefits of the single energy market by limiting the effects of competition and, secondly to take full advantage of energy co-operation with third countries. Development of the gas and electricity interconnections was therefore regarded as integral to the success of the Single European Market, to increase flexibility and to ensure the security of supplies.

By 1996, the two key policy instruments for developing energy TENs – the guidelines for trans-European energy networks (OJL, 1996a) and the favourable context measures (OJL, 1996b) – were adopted. These two policy instruments fulfil Article 129C of the Treaty on European Union which requires the Community to 'establish a series of guidelines covering the objectives, priorities and broad lines of measures envisaged in the sphere of trans-European networks; these guidelines shall identify projects of common interest'. Thus, the Commission was given its first formal role in the development of energy networks.

The guidelines confirmed the objectives of energy TENs as the strengthening of the security of energy supply; to enhance the effective operation of the SEM and to promote economic and social cohesion – in line with the conventional blend of public policy and commercial motivations for public intervention in the energy sectors. In order to achieve these objectives, the guidelines set out priorities for the period up to 2000. These priorities gave formal recognition to ideas expressed in earlier documents. For the electricity networks, priorities included the connection of isolated networks, usually located in less developed regions; the improvement or establishment of interconnections between member states; and the development of links with non-member states elsewhere in Europe and in the Mediterranean region. Gas priorities included the introduction of gas into new regions; the connection of isolated networks to the main European networks; and increases in pipeline and storage facilities to meet both supply and demand concerns relating to gas. Projects to improve the technical management of transnational gas and electricity networks also have priority status.

The guidelines required the EU to identify projects of common interest and proposed a three-stage procedure to this end. Stage one set out the technical definition of projects to be included in energy TENs which were to cover high voltage electricity lines (220 kV or more) and high pressure gas pipelines. The equipment needed for the operation of these lines and pipelines and control centres also came into the definition.

Stage two established the criteria, of which there are three elements, which all projects of common interest must fulfil. A project must come within the above technical definition; it must accord with the objectives and priorities of the guidelines and it must be economically viable. Economic viability covers not only the conventional criteria of financial profitability but also environmental aspects, security of supply and the contribution to economic and social cohesion (OJL, 1996a: Article 6.6).

Stage three identified specific projects of common interest needed to strengthen the networks. Subsequent revisions of the guidelines[7] (OJL, 1996a, 1997b, 1999 and 2003a) have updated and extended the list of projects of common interest in line with the evolution of the markets and networks in the EU-15 and, at the time of writing, proposals brought forward in late 2003 to take account of the 2004 enlargement remain under discussion.

In parallel with the initial passage of the guidelines and in order to accelerate progress in energy TENs, in 1993 the European Council set up a group of personal representative of heads of government under the chairmanship of the then vice-president of the Commission, with a remit to identify priority projects; to eliminate obstacles to the materialisation of projects and to bring forward new regulations to facilitate the projects where appropriate. The main outcome of the resulting Christophersen Report (Group of Personal Representatives, 1995) was the identification of ten priority energy TENs,[8] the purpose of which was to send a clear message of political commitment to the TENs initiative and to give the initiative a focus.

The favourable context measures empowered the Commission to use technical, administrative, legal and financial measures to facilitate completion of projects within a reasonable timescale. Such measures were intended to be used to overcome authorisation problems and to adopt a common approach to rules on health and environmental protection. Ten years after the passage of these measures, it appears that much still needs to be done to realise the objectives of the favourable context measures as cross-border authorisation problems and delays remain the most frequently cited cause of delays in realising network interconnections.

After 1996, the emphasis shifted from policy formulation to policy implementation. During the period 1996–2001, according to a 2001 Commission report on the period, 90 projects of common interest were identified, 24 of which went into operation during this period; 12 were under construction; 22 were in the process of obtaining administrative permits and 32 were at the study stage. Decisive progress had been made on seven of the ten Christophersen projects: all five gas projects came into operation whereas two electricity projects (France-Italy, France-Spain) encountered authorisation difficulties and no final decision was reached on the regulatory framework for connection of the western and eastern Danish networks. EU energy TENs spending came to €123 million, most of which went on feasibility studies (CEC, 2001a).

Based on this five years of experience, the 2001 report also identified new ways in which TENs policy could be fine-tuned and developed. In particular, it called for the introduction of a priority list; a redefinition of TENs priorities which highlighted support for energy market liberalisation, reinforced supply security, integrated renewable energy, incorporated enlargement and gave greater consideration to peripheral and ultraperipheral regions. In addition, it proposed a more general definition of projects of common interest and a less cumbersome process for updating project lists.

These recommendations fed into the first major revision of the TENs guidelines (CEC, 2003a) which came into effect in July 2003. In particular, the guidelines placed even greater emphasis on the completion and operation of the internal energy market, an initiative that was only in its early stages when the original guidelines were drafted but was close to completion, in a legal sense at least, when these new guidelines came into effect. Two common priorities were identified for both gas and electricity networks. The first was to adapt and develop

> the energy networks in support of the operation of the internal energy market and, in particular, solving the problems of bottlenecks (in particular transfrontier ones), congestion and missing links, and taking account of the needs arising from the functioning of the internal market for electricity and natural gas and the enlargement of the European Community (CEC, 2003a: Article 4).

The second reinforced the longstanding social and economic cohesion aspects of TENs – that is, the establishment of energy networks in 'island, isolated, peripheral and ultraperipheral regions'.

Specific priorities relating to electricity networks included integration of renewable energy sources and the promotion of network interoperability within the EU, with accession countries and with other European countries and with the Black Sea and Mediterranean regions. Gas priorities focused on security of energy supplies, emphasising links with third countries and the diversification of natural gas sources and supply routes.

Table 5.2 Trans-European energy networks – priority axes identified in 2003 guidelines

ELECTRICITY NETWORKS	
El.1.	France–Belgium–Netherlands–Germany Electricity network reinforcements to resolve congestion through Benelux
El.2.	Borders of Italy with France, Austria, Slovenia and Switzerland Increasing electricity interconnection capacities
El.3.	France–Spain–Portugal Increased interconnection
El.4.	Greece–Balkans–UCTE system Development of electricity infrastructure to connect Greece to UCTE system
El.5.	UK–Continental and northern Europe Establishing/increasing interconnection and integration of offshore wind energy
El.6.	Ireland–UK Increasing interconnection and integration of offshore wind energy
El.7.	Denmark–Germany–Baltic Ring Increasing interconnection and integration of offshore wind energy
GAS NETWORKS	
NG.1.	UK–Northern Continental Europe (with connections to Baltic Sea Region) – Russia Pipelines connecting some of main sources of European gas supply, improved interoperability of networks and increasing security of supply
NG.2.	Algeria–Spain–Italy–France–Northern Continental Europe New pipelines from Algeria to Europe and increasing network capacities in and between Spain, Italy and France
NG.3.	Caspian Sea–Middle-East–EU New pipeline networks from new sources
NG.4.	LNG terminals in Belgium, France, Spain, Portugal and Italy Diversifying sources of supply and entry points, including LNG links with the transmission grid
NG.5.	Underground storage in Spain, Portugal, Italy, Greece and the Baltic Sea Region

Following on from this, the guidelines identified 12 priority axes for action, seven in the electricity sector and five in the gas sector (see Table 5.2). The guidelines also identify 221 projects of common interest, 122 of which are in the electricity sector and 99 in the gas sector. These projects have been identified in line with the above priorities and additional criteria. Those projects of common interest that fall within the priority axes of Table 5.2 have priority for the limited Community financial assistance for TENs.

The 2004 enlargement rapidly made it necessary to revise the 2003 guidelines and at the end of 2003 (CEC, 2003b) draft new guidelines were brought forward by the European Commission to bring the accession countries fully into the TENs initiative and also to extend TENs to neighbouring countries in line with the new neighbourhood policy and with the need to enhance security of gas supply.

In addition to highlighting the importance of fostering links with the 2004 accession states and neighbouring countries, the proposed guidelines retained the overarching and sectoral specific objectives agreed in the guidelines in 2003 (OJL, 2003a). In addition, the draft includes the following new elements:

1. Finance: given that transit through areas where expensive political risk insurance is a pre-requisite to attract finance (particularly appropriate for gas), the guidelines propose to include the costs of this insurance in the financial regulation for TEN projects;
2. Declaration of European Interest (DEI): that is, the attribution of the highest level of priority to cross-border priority projects having a significant impact on the integration of the networks concerned. The lack of progress on key energy priority projects can be traced to obstacles and variations in national authorisation procedures relating to routeing and environmental consequences. The DEI is intended to help the streamlining of these processes for cross-border priority projects of high European interest involving several member states;
3. European project coordinator: such a co-ordinator or co-ordination team for a priority axis or an individual priority project to improve their implementation would encourage co-operation among users and operators, promote the project among private investors and financial institutions and ensure that the necessary project monitoring is carried out.

Overall, the draft guidelines estimate that €28 billion, most of which will come from the energy network operators, will be required for electricity and gas priority projects during the period 2007–13. The main benefits from such investment are in line with those long perceived to originate from energy TENs, namely continuity and security of supply; lower costs as a result of increased competition, regional development, the integration of the accession countries and neighbours and environmental protection as a result of the increased use of natural gas. However, given the EU's growing import dependency, security of supply is currently accorded the highest importance.

The electricity network

Inter-network co-operation goes back several decades as national incumbents/monopolies interconnected networks to provide assistance in outages and balance supplies in short-term emergencies. In recent years, this co-operation has transformed itself into an attempt at a more complete integration of networks. At the core of this co-operation is the Union for the Cooperation of the Transmission of Electricity (UCTE), founded in 1951 and known as the Union for the Coordination of the Production and Transport of Electricity until 1999. Initially involving the synchronisation of nine networks in Western Europe,[9] UCTE's reach has gradually extended to cover almost 30 countries (Vandenberghe, 2004) with the possibility of even more in the future. By 1992, UCTE's membership, already boosted by that of Greece, Spain and Portugal, had risen to 17 following the break up of Yugoslavia into five countries.

Elsewhere in Europe, the Nordel network involving the Nordic countries was well advanced. By the early 1990s, it had 20 cross-border links with a high voltage submarine link connecting it to Zeeland in Denmark. The Nordel network facilitated the complementary utilisation of the hydro-generation of Norway with the thermal power of Denmark. The UK maintained a separate, slightly smaller system than Nordel, which was linked by a DC interconnector with France.

To the east, the interconnected system of Central and Eastern Europe (IPS), developed from the 1960s, was interconnected with UPS, the interconnected system of the Soviet Union. IPS operated very differently from UCTE but in 1993, the Czech Republic, Hungary, Poland and Slovakia formed Centrel to perform a similar function to UCTE and ultimately to connect with UCTE grids. By 1995, the grids of eastern and western Germany were linked and the grids of Central Europe were synchronised with those of UCTE. Subsequently, Bulgaria was brought into the system and an undersea synchronous connection was made between Spain and three Comelec countries (Algeria, Morocco and Tunisia). Trials have been carried out with Ukraine and feasibility studies are investigating the possibility of links with UPS.

Energy interconnection is about security of supply, among other things and there is large unexploited potential for electricity trade within the EU and indeed between EU members and their neighbours.

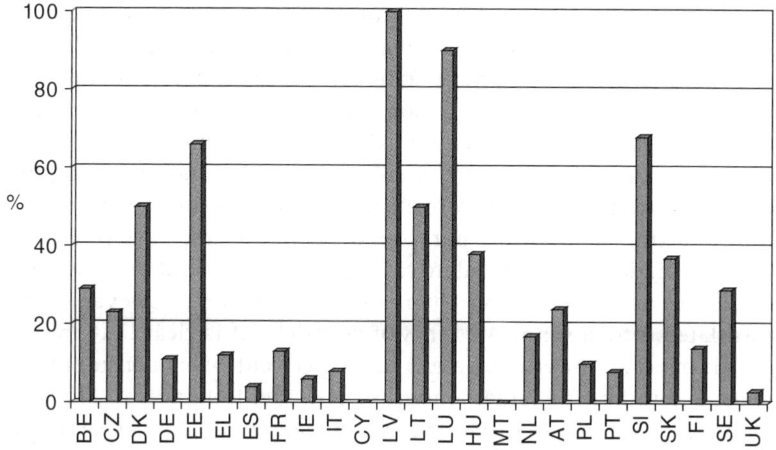

Graph 5.1 Electricity import capacity as a % of installed capacity
Source: CEC (2005).

This trade has hitherto been neglected as a result of the evolution of electricity policy on national lines designed to achieve national self-sufficiency and the incomplete internal energy market. As a consequence, in 2001 the Commission acknowledged that a wider energy market needed not only common rules and standards but also 'adequate infrastructure linking the member countries' (CEC, 2001b). In order to achieve this, the Commission recommended a 10 per cent interconnection target for electricity (CEC, 2001b). At the 2002 Barcelona Council, a target of 'a level of electricity interconnections equivalent to at least 10% of their installed production capacity by 2005' (Presidency Conclusions, 2002) was set. Interconnection initially occurred within the context of a regime of control. Over time, interconnection has shifted towards support for a more liberal regime to provide a foundation for cross-border trade and the creation of an internal energy market.

As Graph 5.1 shows, the degree of import capacity varies across the European Union. Malta and Cyprus, for geographical reasons are not connected at all with other EU countries whilst the interconnectedness of Ireland and the UK remains low. Spain and Portugal, which for a long time operated as completely separate networks, also have a relatively low level of interconnectedness at 4 per cent and 8 per cent respectively. At the other end of the spectrum, Luxembourg is fully integrated and the level of interconnectedness in Nordic countries is high as a result of their closely integrated regional market.

The role played by electricity trade also reveals a differential picture throughout the EU. Clearly, imports and exports of electricity in relation to Cyprus and Malta are non-existent given their absence of interconnector capacity. However, analysis of import and export figures does indicate the extent to which interconnectors are used primarily for importing or exporting purposes or, as in the case of Germany, to keep the trade roughly in balance. Larger countries tend to have lower levels of trade penetration than smaller countries. However, when translated into absolute amounts of electricity it is clear that France, which exports power equivalent to 14 per cent of its consumption, exports significantly larger amounts than, say, Lithuania or Slovenia who export amounts of electricity equivalent to 115 per cent and 52 per cent of their consumption respectively. In short, Graph 5.2, although not useful for absolute comparisons of electricity trade, does indicate the extent to which markets are open in practice and whether the flows are largely inwards or outwards which, in turn, gives some indication of capacity levels and competitiveness. It does not, however, tell the whole picture. The UK, for example, unsurprisingly given its

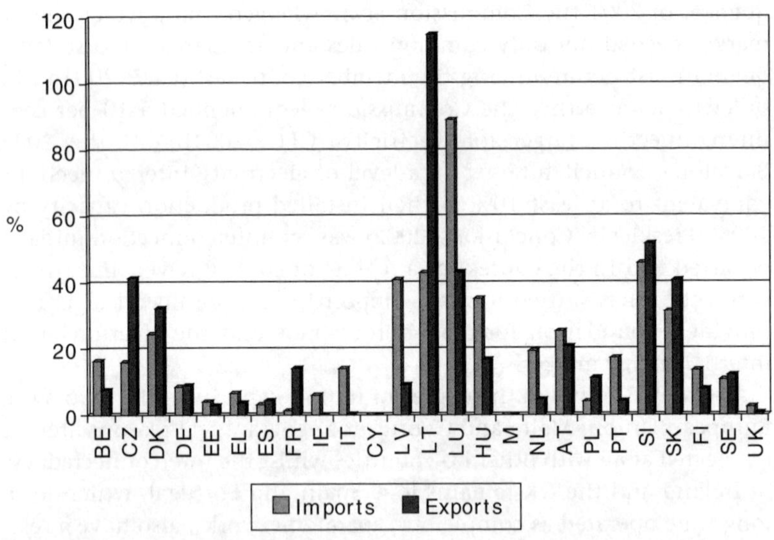

Graph 5.2 Imports and exports as % of electricity consumption in 2004
Source: Derived from Eurostat (2005).

island status trades little electricity in relation to the overall size of its market but its internal market is more competitive than many in continental Europe given the way its liberalisation was carried out.

Electricity trade generally accounts for a greater share of electricity consumption within the Union. In 1989, according to the Commission, electricity imports accounted for 6.6 per cent of the electricity consumption of the EU-12 and exports 5.5 per cent (Johnson and Turner, 1997). The 1995 accession did little to change the figures. By 2003, 9.3 per cent of electricity consumption in the EU-25 was accounted for by both imports and exports. Given that in 2003, electricity imports accounted for 8.4 per cent of electricity consumption and exports 7.5 per cent in the EU-15, 46 per cent of the increased share of imports and 48 per cent of the increased share of exports are attributable to the 2004 accession, implying a much less impressive increase in the share of electricity trade (one possible indicator of the success of liberalisation and the TENs initiative in opening markets), than the bare figures suggest (derived from Eurostat, 2005).

Aggregate figures, however, do mask significant changes in individual countries. In 1989, for example, Ireland had no connections with other networks whilst Portugal's sole electricity exchange was with Spain, amounting to 1 per cent of Portuguese electricity consumption (Johnson and Turner, 1997). By 2004, imports covered 6 per cent of Irish power consumption and 17.2 per cent of Portuguese consumption, signalling greater integration.

Electricity trade is growing in importance for a number of reasons:

1. To meet the essential requirements of security of operation and of mutual assistance – the original motivator for connecting networks;
2. To exploit the potential for synergy – i.e. in order to smooth out load curves and fluctuations in hydroelectric power – one of the prime motivators for the integration of the Nordic networks;
3. To transmit power generated in jointly-financed power units or power transmitted to fulfil long term supply agreements;
4. Increasingly to take advantage of the benefits of the single electricity market – benefits that were not available in 1989 when the TENs initiative was getting underway but which can be expected to come into play more and more as liberalisation advances and utilities exploit new opportunities both within and between markets.

Further development of trade is limited in some parts of the Union by bottlenecks and congestion on the networks, bottlenecks that can have an impact in countries beyond those directly involved. Austria, for example, is centrally located in the enlarged union but is a relatively weak link in the integration of networks and does not fulfil its potential as a transit point for electricity because of alpine conditions and strong local resistance against new power lines which has prevented the construction of new interconnectors to neighbouring countries.

More specifically, the 122 electricity projects of common interest identified in the 2003 guidelines (OJL, 2003a) (and many of which are embedded in the priority axes) are classified in the following way:

1. Developing electricity networks in isolated regions (11 projects);
2. Developing electricity connections between member states needed for the functioning of the internal market (20 projects);
3. Developing electrical connections within member states (55 projects);
4. Developing electricity connections with non-member states (36 projects).

The draft guidelines brought forward at the end of 2003 (CEC, 2003b) left the projects of common interest essentially unchanged. However, they do add the following two new priority axes to those identified in Table 5.2:

- EL.8: Germany-Poland-the Czech Republic-Slovakia-Austria-Hungary-Slovenia: increasing electricity interconnection capacities;[10]
- EL.9: Mediterranean member states – Mediterranean Electricity Ring: increasing electricity interconnection capacities between Mediterranean member states and Morocco-Algeria-Tunisia-Libya-Egypt-Near East Countries-Turkey.

These new axes confirm the priority being given to the extension of the geographical scope of the TENs initiative resulting from accession and from co-operation with neighbouring countries to increase the efficiency, reliability and security of Europe's power grids.

In short, the increased physical interconnection and trade stimulated by energy TENs and European electricity liberalisation have increasingly undermined the control regime by reducing fragmentation and integrating electricity systems. The continued strength of incumbents in some parts of the Union, however, has detracted from this effect.

The gas network

Europe's gas grid is of more recent origin than the electricity network and its configuration is greatly influenced by the location of natural gas production in the European Union which is most developed in the North Sea and the Netherlands with more limited production elsewhere in the European Union. Consumption patterns vary tremendously throughout the Union. Gas comprises 23 per cent of gross inland consumption in the EU-25 but is as high as 46 per cent in the Netherlands; 43 per cent in Hungary; 38 per cent in the UK and 33 per cent in Italy. At the other end of the spectrum, no gas at all is consumed in Cyprus and Malta and gas consumption is 12 per cent or less of gross inland consumption in Greece, Estonia, Poland, Portugal, Slovenia, Finland and Sweden (CEC, 2004c).

Whereas trends in electricity demand growth have tended to be consistent, trends within the gas sector have been driven by a number of forces, thereby placing an array of demands on gas market infrastructure. These forces include:

1. The introduction of gas into new markets: in the mid-1990s, there was no domestic market for gas in Greece and Portugal. Since then, natural gas has been introduced into both markets. In Greece, in the early 2000s, natural gas accounted for 6 per cent of gross inland energy consumption: it had been able to meet its domestic demand by construction of LNG capacity and through the pipeline system from Russia. In Portugal, gas demand had grown to over 10 per cent of gross inland energy consumption over the same period. Like Greece, some of Portugal's gas demand has been met by the construction of LNG capacity and by the construction of a spur from the pipeline coming up from Algeria through Spain and into France;[11]
2. Shifting production patterns: as production begins its decline in the North Sea, and from the UK sector in particular, so the import needs, and with it the infrastructure to carry such imports, needs to be developed;
3. Demand growth: natural gas has steadily increased its share in EU primary energy supply. In 1960, natural gas represented less than 2 per cent of demand: by 1985, it had reached 16 per cent and around 23 per cent in the early 2000s. The Commission estimates gas could account for 30 per cent share of EU energy supply by 2020. Much of this growth in gas demand has come from the increase in gas-fired power generation and co-generation (Europe Commission, 2002).

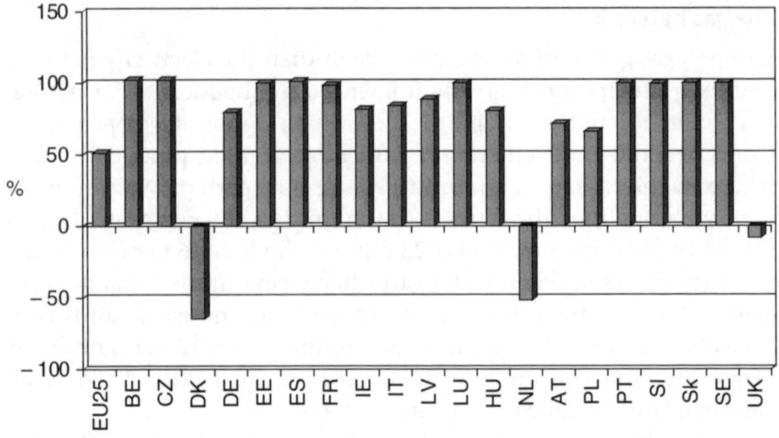

Graph 5.3 Gas import dependency by member state: 2004
Source: CEC, 2004c.

Growing EU gas consumption implies increased EU import dependency. Graph 5.3 demonstrates that ten member states are already at or are approaching 100 per cent reliance on gas imports. The level of import dependency of the others is growing. Given the burgeoning of gas demand in the EU and the poor prospects for increased indigenous gas supply, much of the additional gas needed to fill the supply gap will have to be imported through new pipelines that cross the EU's borders and/or through new LNG facilities.

The key issue for natural gas in Europe is therefore security of supply. Currently the European Union's gas imports originate from Norway, North Africa and Russia (see Table 5.3). Potential exists for future imports from other former republics of the Soviet Union and from the Middle East: indeed deliveries have started to the EU, albeit in small quantities from Qatar, the UAE and Oman and in recent years, LNG from Nigeria has started to find its way into European markets.

Although network development is largely a commercial matter for the gas industry which already, together with co-ordinated stock holding, co-operates in gas transmission to limit expensive, long distance physical transfers, designation of gas projects as a project of common interest within one of the priority axes brings benefits in terms of funding, albeit limited.[12] The 99 gas projects of common interest identified in the 2003 guidelines (OJL, 2003a) are embedded

Table 5.3 2004 EU gas imports by pipeline and LNG (billion cubic metres)

Imports from:	**Norway**	**Algeria**	**Russia**	**Nigeria**	**Others**[1]	**Other member states**
Austria	0.8		6.0			1.0
Belgium & Luxembourg	7.2	2.9	0.2	0.8		10.3
Czech R.	2.6		7.2			0.0
Finland			4.6			0.0
France	14.6	6.7	11.5		0.1	10.9
Germany	26.4		37.7			27.6
Greece		0.6	2.2			0.0
Hungary			9.3			1.6
Ireland						3.7
Italy	7	25.7	21.0	3.8	0.5	9.3
Latvia			1.4			0.0
Lithuania			2.6			0.0
Netherlands	4.3		2.7			6.6
Poland	0.5		7.9			0.7
Portugal		2.3		1.3		0.0
Slovakia			7.3			0.0
Slovenia		0.4	0.6			0.1
Spain	2.2	14.1		4.8	5.5	0.0
Sweden						1.1
UK	9.1					2.3
EU-25						

[1] Includes Libya, Malaysia, Oman, Qatar, the UAE
Source: Derived from BP (2005).

in the priority axes summarised in Table 5.2 and are classified as follows:

1. The introduction of gas into new regions (ten projects);
2. The development of gas connections to meet the needs of the internal market or to strengthen security of supply, including the connection of gas networks (21 projects);
3. Development of more LNG capacity and more gas storage (35 projects);
4. The development of gas supply pipelines (33 projects) – many of which involve bringing more gas into the EU from Russia, former Soviet Republics, the Middle East and North Africa.

As with the electricity networks, the draft guidelines brought forward at the end of 2003 (CEC, 2003b) left the gas projects of common interest essentially unchanged. However, they did propose the following new priority axis to those identified in Table 5.2:

- G.8: Gas pipelines in the Eastern Mediterranean region: the objective is to connect gas from North Africa and the Middle East to the EU *via* Italy and Turkey. Realisation of this priority axis will requires new pipelines in Libya, Egypt, Jordan and Syria.

This new axis confirms the priority being given to supply security and to intensifying energy relations with neighbouring countries.

The liberalisation-infrastructure nexus

The creation of a single energy market was omitted from the original single market plan but plans were quickly brought forward to rectify this and by the late 1980s, the opening of Europe's gas and electricity markets moved onto the EU's agenda. Aided by the interconnections noted above, the IEM was intended to integrate markets to allow operators to enter each others' markets and to shift infrastructure and other aspects of network development in the gas and electricity sectors away from a process of monopoly control.

Opening of these markets proved highly contentious and took time. Helped by the market openings in the UK and the Nordic countries, the crumbling of the view that the power sector was a natural monopoly, at least as far as the services provided over the network were concerned, gave rise to the liberalisation and privatisation that has occurred throughout Europe since the end of the 1980s. Although not uniform across Europe, the tendency in the early days of the electricity supply industry was for development to be pushed by the public sector – but at a local level *via* town councils and municipalities, that is an established control regime. It was only later, when the benefits of scale economies through interconnection of local grids became apparent, that the public role took on a more national perspective and the national state monopolies, the norm throughout much of Europe until relatively recently, emerged. The liberalisation trend has not eliminated the state in this sector but has shifted its focus to regulation and re-regulation. The centrality of electricity to the domestic and commercial sectors makes it likely that this sector will always be under special surveillance by the state. The blackouts in Italy in 2003 have served as a warning that the importance of this particular network goes beyond the purely commercial and that, although the market may have the prime role in operating this sector, there is an overwhelming requirement to keep the power flowing.

The gas sector has gone through similar phases to the electricity sector: the origins of the industry were at the local level. Where the gas industry differs from electricity is in the massive growth of demand for gas, especially for power generation, and the greater necessity for transnational links beyond Europe's borders to secure gas supplies at a time when demand is flourishing whilst indigenous supplies are in terminal decline. This provides major commercial opportunities for multinational enterprises but has also heightened the role of energy diplomacy and the role of public authorities in providing the framework in which these exchanges take place.

The anchoring of energy utilities within a national framework (or in the case of Germany, within a regional framework) is gradually, as a result of nationally and European-driven liberalisation, being superseded by utilities with a more supranational framework. Indeed, in its 2004 document setting out its vision for Europe's electricity market, the European Commission clearly set out its objectives for a European-wide power market.

> The Community is seeking to create a competitive market for electricity for an enlarged European Union, not only where customers have choice of supplier, but also where all unnecessary impediments to cross border exchanges are removed. *Electricity should, as far as possible, flow between member States as easily as it flows within Member States*[13] (CEC, 2004a) (author's italics).

An important part of this drive has been the EU-driven initiatives to liberalise electricity and gas markets throughout the Union. Attempts to develop genuine internal energy markets have been underway since the late 1980s and have encountered serious obstacles along the way. The first major breakthrough was the 1996 Electricity Directive and the 1998 Gas Directive which required the opening of one-third of the electricity and gas markets by 2003 (OJL, 1997a, 1998). In 2003, two further directives were adopted (OJL, 2003b, c) which required full opening of electricity and gas markets for all business customers by 1 July 2004 and for households by 1 July 2007, thereby completing the legal opening of the markets for electricity and gas in Europe.

As Table 5.4 demonstrates in relation to the evolution of European electricity markets, not only did the new legislation requires complete opening of the markets but it also reduced the choice of method available to member states in relation to other aspects of liberalisation, in each case at the expense of the least market-oriented option. For example, when determining whether new power stations are to be built, member states no longer have a choice between an authorisation or a tendering procedure. Following the 2003 Directive, companies apply to build new power stations according to an open and impartial procedure defined by the state rather than bid to build new power stations. In other words, decisions regarding the construction of new generating capacity are to be taken by the corporate sector in line with commercial criteria rather than by the state. Authorisation will be the norm with a tendering procedure only coming into play when there are concerns about supply security or to promote new technologies with an environmental protection dimension. Similarly, in the field of

Table 5.4 Evolution of rules governing Europe's electricity markets

	Prior to 1996	**1996 Directive**	**2003 Directive**
Generation	Monopoly (public/private and/or regional/ national)	Authorisation Tendering	Authorisation*
Transmission & distribution	Monopoly	Regulated TPA Negotiated TPA Single buyer model	Regulated TPA
Unbundling	Not required	Accounts	Legal
Customers	No choice of supplier	Choice for eligible customers (33%)	Choice of supplier for: all commercial customers (2004); all customers (2007)
Regulation	National ministries	Not specified in directive	Independent regulatory authority to be established by each member state

* Tendering still possible if insufficient capacity to ensure supply security or to promote new technology in the cause of environmental protection.
Source: OJL, 1997a and 2003b.

transmission and distribution, regulated TPA, which is the most open and transparent method of granting access to these networks, became the preferred method of opening access to transmission and distribution networks rather than negotiated TPA or the single buyer model.

In practice, Europe's energy markets are still a long way from the European Commission's ideal of allowing electricity to 'flow between Member States as easily as it flows within Member States' (CEC, 2004a). Although less fragmented and more integrated than it was before liberalisation, at best it is possible to talk of 'regional markets' within Europe and, in several parts of the Union, markets remain largely national, albeit a little more porous than before. The European Commission (CEC, 2004a) acknowledges this trend. Figure 5.1 is drawn from the Commission's view of how the regions are shaping up and reflects a non-contentious view of the configuration of the European markets in the mid-2000s. However, the Commission regards regionalisation as a transitory phase which will pass following greater standardisation at European level. In the same way, as neighbouring markets

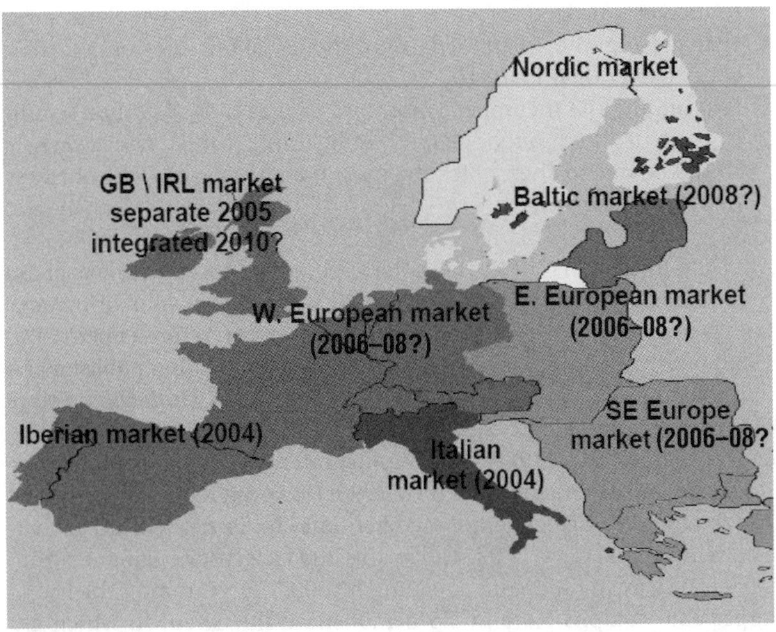

Figure 5.1 Transitional regionalisation of Europe's electricity markets
Source: CEC (2004a).

have become integrated with each other so it is anticipated that the bigger regional entities will gradually follow suit and connect.

Although the achievement of a fully integrated European ESI is logically possible and the trend appears to be moving in that direction, its materialisation is far from guaranteed given the strong forces leading to continued fragmentation. Liberalisation of Europe's energy utilities has taken nearly two decades to achieve and has taken place at various speeds throughout Europe, depending on the level of commitment of individual member states to the free market in this sector. The UK's 1989 Electricity Act, Norway's 1990 Electricity Act[14] and Sweden's Competitive Electricity Market Bill of 1992 kicked off liberalisation in Europe and demonstrated that liberalisation did not necessarily lead to the lights going off. The UK in particular was driven by an ideological commitment to market forces whereas others took longer to convince to go down the liberalisation path. Commonly heard concerns about European plans for liberalisation included:

- A lingering reluctance, especially among countries with few indigenous energy sources, about ceding control of such a strategic sector to market forces;
- Fear among the incumbent monopolies about erosion of their control over their markets;
- Fear among the incumbent monopolies that they would be handicapped by their public service obligations whilst new entrants would be free to cherry pick the most lucrative parts of the business free of such burdens.

After the initial burst of liberalisation at national level, the next phase took place at European level. However, it was far from straightforward. Energy was omitted from the 1985 Single Market White Paper (CEC, 1985) but this was rectified in 1988 when the Commission published an inventory of barriers in the form of a Working Paper *The Internal Energy Market* which formed the basis of subsequent proposals to create a single energy market (CEC, 1988). The original intention was to complete the single gas and electricity markets by 1996. However, it only proved possible to attain the agreement of member states to open one-third of their electricity markets by 2003. It was the 2003 Electricity Directive that paved the way to full market opening by 2007, 11 years after the initial target date. Graphs 5.4 and 5.5 demonstrate the extent to which gas and electricity markets were open in 2004. Unsurprisingly, the 2004 accession states had the furthest to go in terms of liberalisation.

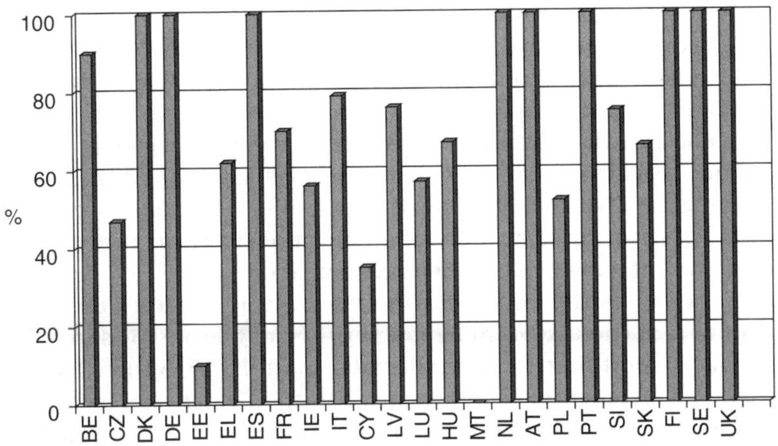

Graph 5.4 Electricity – degree of market opening (2004)
Source: CEC (2005).

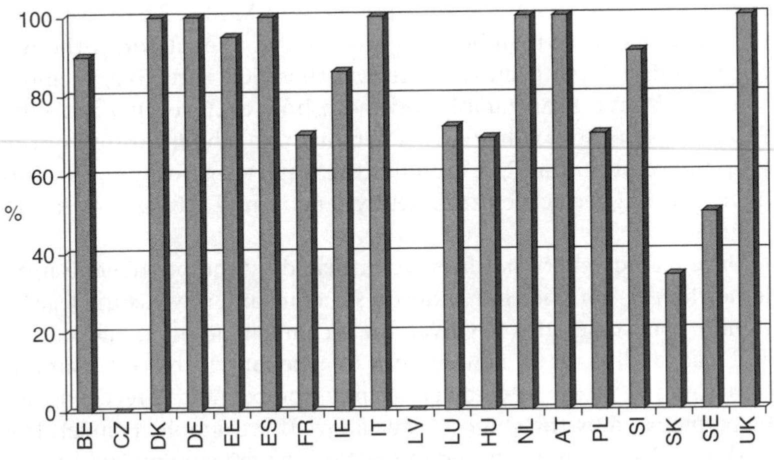

Graph 5.5 Gas – degree of market opening (2004)
Source: CEC (2005).

However, the implications of phased liberalisation, however unavoidable given the variable willingness of member states to open their markets, are that complete market integration, including grid interconnections, is delayed and is replaced by regionalisation. Moreover, although there is a view that early liberalisers obtain first mover advantages, in practice they have experienced enhanced competition within their home market from member states that have not necessarily accorded them the same privileges in their own markets.

EU-driven liberalisation has taken place using directives, a mode of legislation which gives member states a degree of freedom about how they liberalise their markets providing they meet the overall objective (in the case of the 2003 Electricity Directive – 100 per cent market opening by 2007). In general, the flexibility of directives enables member states to adapt European legislation to their own specific environment and avoids excessive centralisation.

In the case of the Electricity Directives, member states were given a number of choices regarding their method of liberalisation in order to gain support from reluctant liberalisers. In the 1996 Directive, for example, member states were allowed to choose between tendering and authorisation methodologies when determining who would be allowed to build new power stations. In transmission and distribution, in order to obtain the support of France in particular, member states were allowed to choose between regulated third party access (TPA), negotiated TPA and the Single Buyer Model. There was also no requirement for member states to appoint a regulator. Although these choices have been restricted by the 2003 Directive, member states still have considerable leeway in how they comply with the Directive. On the one hand, this demonstrates the flexibility of the EU. On the other hand, it muddies the waters regarding the attainment of a single electricity market by sanctioning different rules and regimes.

Moreover, diversity has been accentuated by the pre-liberalisation legacy in terms of tradition, industry structure and pre-existing legislation. In some cases, this has given the incumbent an inbuilt advantage and made it difficult for new entrants to penetrate the market. This can sometimes be exacerbated by geography (for example, Greece) or can be perpetuated by the choices made above (for example, France). The upshot is that, both in law and in practice, the preparations for liberalisation and liberalisation itself have largely been left to member states with the result that market opportunities remain uneven throughout Europe. Even where the market is completely open, as in Germany, the oligopolistic nature of the market has reduced the attractiveness of the market for newcomers.

The decentralisation of Europe has made the issue of EU-level regulation highly delicate. It was only in July 2004 that Germany finally acquiesced to the requirements of the 2003 Directive and moved away from a system of self-regulation to that of an independent national regulator. However, significant national variations remain in regulation, competition law and trading regimes. This, plus the absence of a coherent cross-border trading system and ongoing authorisation delays inhibit further development of interconnections. There is a view that the attainment of a genuine single electricity market in line with the Commission's medium-term vision requires a more radical innovation in the form of a European regulator. Initiatives like the Florence Forum and the creation of the European Council of Regulators represent efforts to achieve greater uniformity in regulation on a more voluntary level but this ideal remains a long way from becoming reality.

The ongoing fragmentation and differential policy and regulation throughout Europe's gas and electricity markets combined with different authorisation requirements for new links combine to act as continuing obstacles to the development of integrated and interconnected European gas and electricity networks, thereby inhibiting the realisation of the full benefits of liberalisation. However, liberalisation is an ongoing process and, as anomalies arising from uneven liberalisation come to light, so further movement towards ending these anomalies can be expected. Even more to the point, creation of the single gas and electricity markets is a process and not an event and the reaction of the corporate sector to the shifting regulatory landscape takes time to unfold.

Overall, the European gas and electricity utility sectors have moved away from the traditional out-and-out control regime. However, reaction to the easing of control has varied from member state to member state. In some member states, despite the EU liberalisation initiatives, incumbents continue to dominate their national markets with the approval and even connivance of their governments. At the other extreme, liberalisation has been taken fully on board and markets operate in a fully competitive manner. In the mid-2000s, the reality for most markets is somewhere in between.

The corporate response

Liberalisation and the possibilities it creates have transformed the business environment in which gas and electricity utilities operate and have thus introduced new strategic considerations and tools (Dyner and Larsen, 2001; Sioshansi, 2001; Skytte, 1999). Prior to liberalisation, competitiveness was not an issue: rather the main concern was to guarantee supply and to satisfy other public service obligations. This generated the regime of control. The senior management of the energy utilities were dominated by engineers whose main concern generally was to attain the best technical solution to problems rather than to seek the most cost-effective solution. Traditional planning methods incorporating large-scale hard models, simulation and forecast of the overall markets were used to prepare for the future. In some senses, the industries took on some of the characteristics of central planning utilised by the Soviet blocs as they strove to meet targets and objectives set by governments[15] rather than by responding to the market, which was captive.

In a more liberal scenario, utilities must become more responsive to changes in the environment, which occur more quickly. They also have to take into account factors that were previously irrelevant: the only competition these utilities encountered in bygone days was competition between gas and electricity in limited applications. There was no competition within individual sectors and thus little heed was paid to issues of consumer choice or the creation and management of brands. Key stakeholders were the government, taxpayers and employees. Private shareholders in many, albeit not all, European countries did not exist and thus concerns about shareholder value were also absent. Since liberalisation, these factors have been turned on their head, requiring new approaches to all types of commercial risk and the utilisation of business models and strategic tools that are commonplace in the rest of the business world. This shift from monopoly to market requires a complete change of corporate culture and structural reorganisation – a process which is more developed in some countries than others.

At the onset of European-wide liberalisation, the majority of power and gas utilities were primarily owned by the public sector, either in the form of a nationally integrated system like France, Greece, Ireland and Italy or with networks and systems controlled by state, regional or municipal authorities with little private sector participation like the Netherlands and Denmark. The UK and the Nordic companies had liberalised their power utilities (and in the case of the UK its gas utilities) before the 1996 Electricity Directive. Portugal and Spain were in the process of transition from a traditional monopolistic/oligopolistic system towards a more competitive system and there were other anomalous systems like Luxembourg and Ireland that were neither integrated nor liberalised but were in a unique position, because of its size in the case of Luxembourg and its island situation in the case of Ireland. The 2004 accession states were in the process of transition as a result of their general transformation and have become fully integrated into the European liberalisation process.

Whatever, the exact timing of their liberalisation, however, it is fair to say that control was the norm for almost all energy utilities in Europe until the final decade of the twentieth century. The state both owned, in one form or another, and or controlled the strategic direction of the industry. The industry was often subject to non-commercial decisions imposed upon them by the state for other, often political, reasons.[16]

Given the diversity in terms of timing, extent and form with which gas and power liberalisation was introduced in Europe, the move away from control occurred differentially and reflected the specific circumstances and power relationships within each submarket. In general, successful liberalisation requires competition in generation and supply; access to networks and, ideally cross-border interconnections to reduce market power and further increase competition. Markets in the EU, however, remain fragmented, in part because the process of liberalisation is still underway and the industrial restructuring begun by this is in its infancy: in part because of strategic choices made by key players in major markets and in part because bottlenecks and missing links remain in transmission networks.

In most cases, when liberalisation occurred, member states did not break up the dominant supplier to reduce the market power of incumbents with the result that concentration remains high in many markets (as demonstrated by Table 5.5 and Graph 5.6) and entry for newcomers is difficult (the UK has been an exception to this – see below). Although care needs to be taken in interpreting Table 5.5, it does provide some indicator of market opening and the degree of competition in power markets in various member states. Horizontal integration, or market concentration, results from the concentration of ownership or control of one or more of the activities in the value chain. It enables firms to withhold generating or production capacity to manipulate the market or abuse their dominant position in the operation of transmission systems. With one producer owning 65 per cent or more of generating capacity, Table 5.5 indicates high levels of market concentration in power generation in the Czech Republic, France, Greece, Ireland, Latvia, Portugal, Slovenia and Slovakia. In Belgium, Germany,[17] the Netherlands, Hungary, Sweden, the UK and Poland, the largest generator owns 30 per cent or less of generating capacity, implying a reduced ability to manipulate the market. In supply, the three largest suppliers control over 80 per cent of supply in Belgium, France, Greece, Ireland, Latvia, Lithuania, Luxembourg, the Netherlands, Portugal and Spain. Moreover, the market share of foreign

Table 5.5 Degree of market concentration in Europe's electricity sector

	Generation		**Suppliers**		
	Largest producer by share of capacity (%)	**3 largest producers by share of capacity (%)**	**Number with market share > 5%**	**Share of top 3 suppliers (%)**	**Market share of foreign owned suppliers (%)**
Austria	45	75	4	67	2
Belgium	30	70	2	90	10
Czech R.	65	75			
Denmark			5	67	Not known
Estonia			1	?	3
Finland			6	30	25
France	85	95	1	88	9
Germany	30	70	3	50	20
Greece	100	100	1	100	0
Hungary	30	65	7	56	97
Ireland	85	90	4	88	12
Italy	55	75	6	35	Not known
Latvia	95	100	1	99	0
Lithuania	50	80	1	100	0
Luxembourg	n.a	n.a	2	100	0
Netherlands	25	80	3	88	18
Poland	15	35	3	32	17
Portugal	65	80	3	99	33
Slovakia	75	85	4	84	28
Slovenia	70	95	6	71	20
Spain	40	80	5	85	8
Sweden	15	40	4	70	39
UK	20	40	6	60	50

Source: Derived from CEC (2005).

suppliers in most of these countries is amongst the lowest in the EU. This implies limited options for switching suppliers – one of the key indicators of a more competitive market.

Similarly Graph 5.6 shows the extent to which, despite the liberalisation of European gas markets that has taken place to date, many European markets remain almost completely dominated by one gas company. Indeed the UK, Spain and Ireland are the only member states where the largest company controls less than 40 per cent of available gas. In seven member states, the largest companies control 100 per cent, or approaching 100 per cent of available gas.

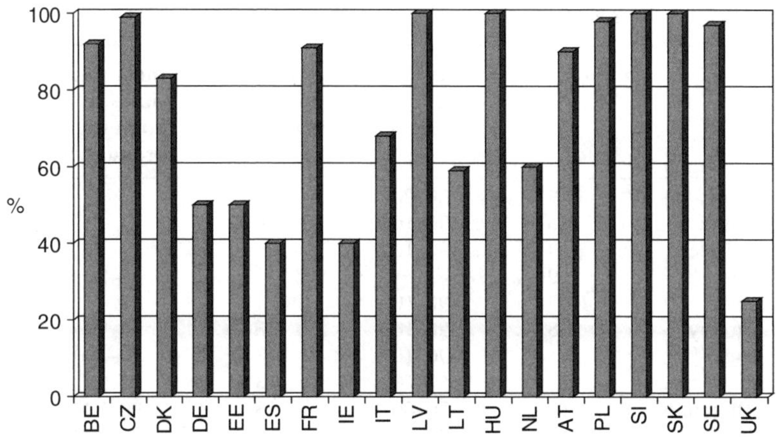

Graph 5.6 Share of available gas controlled by largest company
Source: Derived from European Commission (2005).

In other words, despite liberalisation, gas markets remain concentrated and dominated by incumbents. The gas consumed in Europe comes from indigenous production, imports and from trading hubs. As Table 5.6 shows, and in the process confirming the picture conveyed by Graph 5.6, incumbents supply the majority of gas through long-term contracts based on imports or domestic production. Many of these contracts pre-date liberalisation and were frequently linked to specific projects to offset some of the risk in developing the infrastructure needed to secure gas supplies. Only in the UK, one of the few markets where full legal unbundling has taken place, has the role of the incumbent been cut back. The situation is summed up in the preliminary report on the Competition DG's inquiry into the Energy Sector which stated:

> At the wholesale level of the gas supply chain EU liberalisation has not, so far, significantly changed the market structure. The high level of concentration which existed in most national markets at liberalisation largely remains. This is true at both the national wholesale and retail levels for most countries ... In many national markets no liquid wholesale market has emerged and traded markets (gas hubs) represent a minor part of gas supply (CEC, 2006: 28).

Vertical integration involves the control by one firm of the stages of the utility supply chain from generation/production, through transmission,

Table 5.6 Market concentration in the wholesale gas market – 2004

	Imports – bcm	**Incumbent share of imports (%)**	**Domestic production – bcm**	**Incumbent share of domestic production (%)**
Austria	9	80–90	2	–
Belgium	16	90–100	0	–
Czech Rep.	9	90–100	<1	–
Denmark	0	–	10	80–90
France	49	90–100	1	–
Germany	88	90–100	18	80–90
Hungary	11	90–100	3	90–100
Italy	67	60–70	13	80–90
Netherlands	18	50–60	73	90–100
Poland	10	90–100	5	90–100
Slovakia	7	90–100	<1	–
UK	13	20–30	105	40–50

Source: CEC, 2006.

distribution and eventual supply to the consumer. Vertical integration is a widespread tendency throughout European power and gas sectors, a process which, despite some unbundling and strengthening of regulation, dilutes the impact of liberalisation and increases the market power of incumbents. This makes possible many types of anti-competitive behaviour, including cross-subsidy of activities or making access to networks for new entrants difficult or even impossible. It also reduces the incentive to trade on wholesale markets, rendering them illiquid and thus reducing incentives for new entrants. Unbundling (namely, the separation of production and supply from transmission and distribution) and regulation to ensure the transmission grid is open to all in a non-discriminatory manner can help prevent this. Even where unbundling has been carried out in line with the relevant directives, the 2006 inquiry into competition in the energy sectors raises the issue that the rules do not provide a safeguard against vertically integrated companies favouring their own supply affiliates at the expense of their competitors (CEC, 2006).

From the perspective of individual utilities, however, vertical integration is attractive and offers synergistic possibilities as a result of collaboration between different segments of the value chain and allows for compensation for market fluctuations at one stage of the value chain by fluctuation elsewhere. Several of Europe's larger utilities have gone

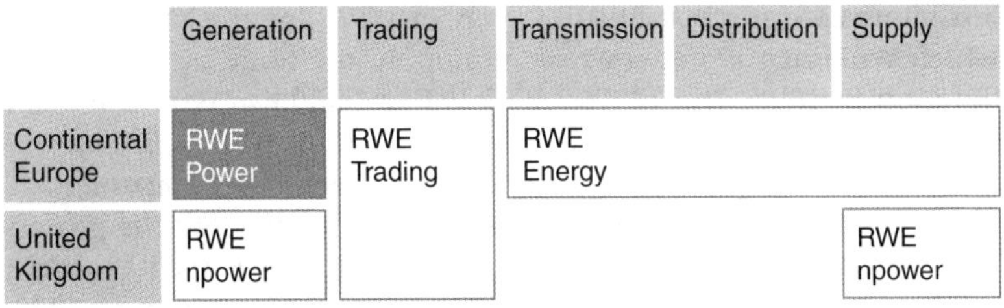

Figure 5.2 Example of a vertically integrated utility
Source: RWE (2005).

down the vertical integration path. Figure 5.2 for example, shows how RWE (Europe's third biggest power utility) has taken vertical integration on board. Moreover, the integration of trading into the RWE value chain provides the company with additional safeguards in managing risk.

Vertical integration is also the watchword among utilities involved in gas. E.ON, a large German utility involved in both electricity and gas following its acquisition of Ruhrgas in 2003, also maintains vertical integration at the core of its strategy. In UK and Nordic markets, for example, it is involved in production, has a stake in the Bacton-Zeebrugge Interconnector and is an important player in both the UK and Nordic retail markets. E.ON is also getting involved in the Russian upstream gas[18] industry, in the development of the gas pipelines to bring gas to Europe and is searching for further joint ventures in downstream Europe. Both RWE and E.ON are heavily involved in both the gas and power sectors, facilitating the exploitation of synergies between them, especially as a result of the growth of gas as a source of power generation within Europe.

Even where liberalisation involved vertical disintegration of markets, over time there is a tendency for reintegration to occur as a result of the inevitable restructuring that follows such a profound shake up of the market. Reintegration is gradually occurring in the UK market, for example. As one of the first markets to liberalise and to genuinely open itself to competition, the UK market provides a good case study of how markets can respond to liberalisation. Prior to privatisation in 1990, the UK's previously monolithic ESI was broken into three regions (England & Wales, Scotland and Northern Ireland). In England and Wales, the Central Electricity Generating Board was broken down into four companies: two generators – National Power and Powergen; Nuclear Electric, which remained in the public sector until it was privatised as British Energy in 1996; and the National Grid which remained a state-owned monopoly for some years until its flotation in 1995. The 12 Area Boards that had been responsible for supply and distribution in the days of state ownership became 12 independent and privately owned Regional Electricity Companies.

The UK phased in competition over a number of years: the process was only completed by 1998 when all consumers gained the right to choose their electricity supplier. In the years leading up to 1998, there was a flurry of mergers and acquisitions. By 1999, five RECs were owned by US companies, although these later withdrew from the market, EdF had bought London Electricity; Scottish Power had bought Manweb and two RECs (Norweb and South Wales Electricity) had merged with local water utilities and Southern Electric had acquired Southern Water. Southern also obtained a licence to become a distributor of natural gas, making it the forerunner of the many companies that became dual fuel providers within a short space of time to take advantage of the introduction of competition into gas markets. Powergen and National Power, at this stage, maintained their dominance over generation but investment in independent power producers was also thriving.

By 2005, the disintegrated post-privatisation structure of the UK industry was difficult to detect. In 2002, the Lattice Group, formerly Transco – the former transmission business of British Gas, had merged with the National Grid. EdF owned three RECs (Eastern, London and South East Electricity) and supplied one-quarter of the UK's population with electricity and generated 7 per cent of the UK's electricity. Indeed, all the RECs had generating interests and Powergen and National Power had both been absorbed by foreign companies that had already established a major presence in distribution and supply. In 1995, the UK government sold its remaining 40 per cent stake in Powergen. By 1998, Powergen had embarked upon a strategy of vertical integration by acquiring East Midlands Electricity. Shortly afterwards, it began to sell electricity and gas to domestic consumers over the Internet and became involved in telecommunications. In 2002, E.ON completed the acquisition of Powergen and in 2003, the company acquired Midlands Electricity which, together with East Midlands Electricity, was subsequently rebranded as Central Networks. In July 2004, Powergen UK was rebranded as E.ON UK.

National Power has undergone a similar process. In 1999, it acquired Caltorex and its 500,000 gas customers. Later, the same year, it acquired the electricity supply business of Midlands Electricity and created npower to represent its retail business. Later in 1999, Npower announced a demerger which resulted in the formation of Innogy to manage its UK power and generation businesses and International Power. The following year, Innogy acquired Yorkshire Electricity and in 2002 exchanged the distribution side of Yorkshire Electricity for the customer base of Northern Electric and Gas. In May 2002, Germany's RWE acquired Innogy and was later renamed RWE npower. By this stage, the original National Power was barely detectable in the new company which had become a vertically integrated subsidiary of one of the world's biggest multi-utilities.

Outside the big three (EdF, RWE and E.ON), there had been some further consolidation within the industry. South Wales and South West had come together in Western Power Distribution and Southern and Scottish Power had come together in SSE Power Distribution. United Utilities, the product of the much earlier merger of North West Water and Manweb continue to push forward their multi-utility strategy.

Overall, the UK market has made a significant move towards a liberal regime. Despite the massive restructuring that has taken place and the growing influence of EdF, RWE and E.ON, the market remains highly competitive as testified by the degree of customer switching that takes place and the wide range of choice of suppliers available to consumers. The role of the state, for the time being at least has become much more passive. Initially the regulator frequently intervened in the marketplace to control prices and required the two main generators to divest themselves of generating units to increase competition. As the liberalised industry has matured, regulation has become more arm's length and passive. It is the competition authorities of the UK and the European Union that increasingly act as the guardians of competition by scrutinising the competitive effects of proposed restructuring.

Given the critical contribution of energy to the UK economy, a passive approach will become more active, however, if the UK's energy supply is perceived to be under threat. For example, concerns about greenhouse gases and possible power shortages have already revived the debate over nuclear power. The construction of nuclear power stations in the UK will require a political decision and movement towards a more active policy. As domestic production falls, gas will also become an issue in the UK: in 2006, rumours about a potential bid for Centrica by Russia's Gazprom's intention to enter the UK market have sparked a debate about gas supply security in the UK, spurred on by events in Ukraine in January 2006. The UK government has said that it would not attempt to block any such moves by Gazprom but if such a move does proceed, it will be highly controversial and will further politicise energy issues in the UK.

Liberalisation increases competition by encouraging market entry. Market entry can occur either by investment in a new market or by the physical connection or integration of markets. The former depends on the willingness of governments to welcome new players, which is not always the case (see below) and on the absence of entry barriers raised by incumbents. This type of market integration has happened in separate parts of the European market. Utilities from France, Germany and to an extent, the Nordic countries, have broken out from their domestic base and established positions in the UK, Central and Eastern Europe, etc. However, this form of market integration is not really European market integration but rather the introduction of competition into specific segments of the overall European market.

European market integration requires greater cross-border trade, which in return requires greater cross-border interconnections, to make this a reality. Energy TENs suggests potential axes and is predicated on private financing of the networks. In order for the market to provide a clean steer for infrastructure development, appropriate incentives need to be in place. Market concentration, vertical integration and the thinness of wholesale markets resulting from power purchasing agreements and long-term gas contracts, many of which precede European energy liberalisation do not provide these incentives. For example, a vertically integrated network operator in a heavily concentrated sector has no incentive to expand and/or connect its network if the main beneficiaries will be its competitors. The Competition Directorate's preliminary findings in its Energy Sector inquiry confirm the inability of imports to play a full role in opening the market (CEC, 2006). In both gas and electricity markets, there was insufficient cross-border capacity to meet demand and high levels of congestion and inefficient congestion management methods on key routes. In addition, in electricity significant differences in the rules managing different networks detracted from market integration and long-term cross-border capacity reservations were often granted under discriminatory conditions. In gas, incumbents have proved particularly successful in securing transit capacity on existing key routes and on expanded transit lines because of their long-term contracts, thereby making it extremely difficult for new entrants. Overall, the power of incumbents both within their markets and in terms of their control over the physical means of access makes it difficult for new entrants to gain access to many national markets, thereby limiting interconnectedness and the creation of single energy markets.

If all markets had the same liberal approach as the UK, the EU's utility markets would already resemble the Commission's vision for the EU energy markets rather than the regional framework that applies in the mid-2000s. In practice, asymmetric strategies as alluded to by Bonardi (see Chapter 1) have taken hold in some member states. Some of the EU's most aggressive utilities in terms of developing their portfolio and activities outside their home country are from countries where competition is not established in the same way as it has in the United Kingdom.[19] EdF, E.ON and RWE, for example, are among the largest utilities in Europe with involvement in a number of European markets outside their home territory. However, the markets of France and Germany, despite the 100 per cent legal opening of the latter, show limited signs of competition in their domestic markets either in terms of customer switching or in terms of the number of foreign suppliers in their market. At present, these markets operate under a corporate regime: the domestic markets remain dominated by domestic producers (or in the case of Germany, regions are dominated by local suppliers) and state efforts to introduce competition so far, have been limited and/or ineffectual.

The gas and power markets continue to evolve as liberalisation unfolds. Challenges to those countries whose policy stance has facilitated the adoption of an asymmetric strategy by their main utilities can be expected from other countries whose access to their markets is denied or limited whilst allowing their utilities access to their markets. Legal challenges at EU level and completion of the privatisation of EdF and GdF may help to leverage *de facto* as well as *de jure* opening. However, it is possible that by the time these markets are prised open, if this indeed happens, utilities from these countries that have been able to enter other markets whilst benefiting from a protected home market will have built such a strong position that they will have an inbuilt advantage in pushing on to become a genuine pan-European utility.

This trend is intensifying and the preference for protecting the home market is becoming more explicit. This results from a general growing concern about energy security and specific concerns about electricity supply (not helped by the 2003 Italian blackouts) and the reliability of gas imports (not helped by the January 2006 stand-off between Russia and Ukraine which disrupted supplies going further west). According to the rationale for TENs and energy liberalisation, interconnectedness and market integration should help overall supply security but in practice member states, or at least some of them, appear to be taking the stance of developing national champions, not only to help them compete in a broader market but more specifically for strategic reasons and, one suspects, to give them greater capability to operate unilaterally in times of crisis.

In the first few months of 2006 alone, there have been two examples of states taking a much more active policy stance on energy matters. In both cases, the states have emphasised the strategic nature of energy and the need to maintain national control over their energy resources.

Conclusion

Given the height of the barriers facing gas and electricity utilities prior to liberalisation, it is not surprising that, once the barriers began to fall, these companies underwent some of the most extensive restructuring of all sectors affected by the single market. In the mid-2000s, both Europe's gas and electricity sectors are in a transition phase, a transition that is likely to last for some time given that liberalisation itself is not complete until 2007; that liberalisation has been implemented more thoroughly and with more enthusiasm in some parts of the EU than in others and that, in any case, the response to liberalisation and other related policies takes time to unfold and frequently goes through several iterations.

The upshot of the above is that Europe's utility markets, although more connected than in the past, remain more fragmented than full implementation of Europe's liberalisation policies would suggest. This resistance to a shift to a more liberal regime has been reinforced by the response of some utilities to the policy changes – namely, the strategies of vertical and horizontal integration adopted by some of Europe's largest utilities. In several cases, these utilities have been helped by the continuation of many aspects of a control regime in their domestic market (thereby enabling them as incumbents to develop a strong domestic position that is difficult for utilities from other member states to challenge) whilst themselves being able to take advantage of more liberal regimes elsewhere in the EU.

In short, an unintended consequence of this uneven liberalisation has been the creation of corporate regimes in some parts of the EU. This market consolidation is helping create pan-European energy players but the process is not that envisaged by the architects of the single gas and electricity markets nor is the outcome in terms of pluralistic market development. For the most part, the process has also not supported the creation of an environment that is propitious for the development of energy TENs.

References

BP (2005) *Statistical Review of World Energy*, London: BP.

Commission of the European Communities (CEC) (1985) *Completing the Internal Market (White Paper* from the EC Commission to the European Council), COM(85) 310.

Commission of the European Communities (CEC) (1988) *The Internal Energy Market*, COM(88) 238.

Commission of the European Communities (CEC) (1990) *Towards Trans-European Networks for a Community Action Programme*, COM(90) 585 final of 10 December 1990.

Commission of the European Communities (CEC) (1992a) *Proposal for a Council Regulation (EEC) introducing a declaration of European interest to facilitate the establishment of trans-European networks in the electricity and natural gas domain*, COM(92) 15–24 February 1992.

Commission of the European Communities (CEC) (1992b) *Electricity and Natural Gas Transmission Infrastructures in the Community – Communication from the Commission to the Council*, SEC (92) 553 final of 27 March 1992.

Commission of the European Communities (CEC) (2001a) *(Draft) Report from the Commission to the Council, the European Parliament, the Economic and Social Committee and the Committee of the Regions on the implementation of the guidelines for Trans-European Energy Networks in the period 1996–2001*, Brussels, 14 December 2001.

Commission of the European Communities (CEC) (2001b) *Communication from the Commission to the European Parliament and Council: European energy infrastructure*, COM(2001) 775 of 20 December 2001.

Commission of the European Communities (CEC) (2002) *Communication from the Commission to the European Parliament and Council – the internal market in energy: coordinated measures on the security of energy supply*, COM(2002) 488.

Commission of the European Communities (CEC) (2003a) *Communication on Energy Policy and Neighbouring EU countries*, COM(2003) 262 of 13 May 2003.

Commission of the European Communities (CEC) (2003b) *Commission Staff Working Paper: Decision of the European Parliament and of the Council laying down guidelines for trans-European energy networks and repealing Decisions No 96/391/EC and 1229/2003/EC*, COM(2003) 742 final.

Commission of the European Communities (CEC) (2004a) *Medium Term Vision for the Internal Electricity Market*, DG Energy and Transport Working Paper, Brussels, 1 March 2004.

Commission of the European Communities: Directorate General of Energy and Transport (CEC) (2004b) *2003 – Annual Energy and Transport Review.*

Commission of the European Communities (CEC) (2004c) *Energy and Transport in Figures: Statistical Pocketbook 2004.*

Commission of the European Communities (CEC) (2005) *Commission Staff Working Document: Technical Annexes to the Report from the Commission on the*

Implementation of the Gas and Electricity Internal Market, SEC (2004) 1720, Brussels, 5 January 2005.

Commission of the European Communities (CEC) – Competition DG (2006) *Energy Sector Inquiry – Draft Preliminary Report.*

Dyner, I. and Larsen, E. (2001) 'From planning to strategy in the electricity industry', *Energy Policy,* 29: 1145–54.

Eurostat (2005) *Electricity Statistics: Provisional data for 2004,* Statistics in focus; Environment and Energy – 5/2005, Luxembourg: Eurostat.

Group of Personal Representatives of the Heads of State or Government (1995) *Trans-European Networks,* Luxembourg: Office for Official Publications of the European Communities, colloquially referred to as the 'Christophersen Report'.

Johnson, D. and Turner, C. (1997) *Trans-European Networks: the Political Economy of Integrating Europe's Infrastructure,* Basingstoke: Macmillan Press.

Official Journal (1976) *Council Regulation (EEC) No 1215/76 amending Council Regulation (EEC) No. 1056/72 (OJL 120 15.5.72, p.7),* OJL 140 of 18 May 1976, p.1.

Official Journal (1996a) *Decision 1254/96/EC of the European Parliament and the Council of 5 June 1996 laying down a series of guidelines relating to the trans-European networks in the energy sector,* OJL 161 of 29 June 1996, pp.147–53.

Official Journal (1996b) *Decision 391/96/EC of 28 March 1996 determining a set of actions with a view to establishing a more favourable context for the development of the trans- networks in the energy sector,* OJL 161 of 29 June 1996, pp.154–5.

Official Journal (1997a) *Directive 96/92/EC of the European Parliament and of the Council of 19 December 1996 concerning common rules for the internal market in electricity,* OJL 27 of 30 January 1997, pp.20–9.

Official Journal (1997b) *Decision 1047/97/EC of the European Parliament and the Council of 29 May 1997 amending Decision No 1254/96/EC laying down a series of guidelines for trans-European energy networks,* OJL 152 of 11 June 1997, pp.12–14.

Official Journal (1998) *Directive 98/30/EC of the European Parliament and of the Council of 22 June 1998 concerning common rules for the market in natural gas,* OJL 204 of 21 July 1998, pp.1–12.

Official Journal (1999) *Decision 1741/99/EC of the European Parliament and the Council of 29 July 1999 amending Decision No 1254/96/EC laying down a series of guidelines for trans-European energy networks,* OJL 152 of 6 August 1999, p.1.

Official Journal (2003a) *Decision 1223/2003/EC of the European Parliament and the Council of 26 June 2003 laying down a series of guidelines for trans-European energy networks and repealing Decision 1254/96/EC,* OJL 176 of 15 July 2003, pp.11–28.

Official Journal (2003b) *Directive 2003/54/EC of the European Parliament and the Council of 26 June 2003 concerning common rules for the internal electricity market,* OJL 176 of 15 July 2003.

Official Journal (2003c) *Directive 2003/54/EC of the European Parliament and the Council of 26 June 2003 concerning common rules for the internal gas market,* OJL 176 of 15 July 2003.

Presidency Conclusions (2002) *Barcelona European Council,* 15 and 16 March 2002.

RWE (2005) *RWE Facts and Figures* (updated June 2005).

Sioshansi, F. (2001) 'Viewpoint: opportunities and perils of the newly liberalized European electricity markets', *Energy Policy*, 29: 419–27.
Skytte, K. (1999) 'Market imperfections on the power markets in northern Europe: a survey paper', *Energy Policy*, 27: 25–32.
Vandenberghe, F. (2004) 'Development of Interconnections and Reliability Standards', in Bielecki, J. and Desta, M. (eds) *Electricity Trade in Europe: Review of the Economic and Regulatory Changes*, Alphen aan de Rijn, NL: Kluwer.

6
Themes in the Development of the European Network Economy

The network sectors are key to Europe's position in the modern global economy and, as one leg of the 'triad' of regions that dominates the international business environment, Europe's policy-corporate strategy interface is potentially highly significant, both within the continent and beyond. The conceptual framework offered in Chapter 1 seeks to identify the nature of the policy-strategy interface which, over the past two decades has altered in line with more passive industrial policies and decreasing state involvement.

The European Commission has promoted a process of regime shift (alongside and complementing that occurring at the nation state level) whereby it sought to remove network fragmentation by creating a network of networks. This process was to be realised through and supported by market processes. The actions of private operators, so the Commission hoped, would drive the harmonious development of networks across Europe and offer market-driven solutions to the key problems faced by the respective network industries. These networks do not merely reflect private concerns: each also performs socioeconomic functions which secure the rights and needs of humanity as well as offering commercial support to business. Thus, the networks have to address broad social as well as private commercial needs and the crux of the policy-corporate strategy interface is the extent to which policy makers have faith in markets and the extent to which markets can fulfil these needs. These are represented by the mobility gap, security of energy supply and the emergent information society.

Inevitably, this process of regime shift has taken place against a background of technological and commercial change. This has created new challenges as well as new strategies and business models and resulted in potentially new solutions to key policy problems. The maturity of competition has led to new types of services and to increased competition within specific sectors that has allowed some operators to alter the way in which infrastructure is viewed as a transport and distribution mechanism. These developments have just as much potential to alter the planning and development of infrastructure as many of the issues raised above.

Reflections on the policy/strategy framework

What marks the TENs sectors out from many other sectors is the combination of the essential support they give to the rest of the economy and the non-commercial, politically-related factors which cause nation states to continue to intervene in these sectors to varying degrees, even after they have relinquished key aspects of their control. It is the contribution of the TENs sectors to the basic economic functions of communications, mobility and production, which oil the wheels of trade and commerce, that confers particular economic importance upon them. Not only does the success or otherwise of these key sectors have serious implications for the competitiveness of European economies but their impact spills over into other spheres, giving each sector a significant political dimension. It is this ongoing tension between the commercial and the political which has been a recurring theme throughout this book and which the basic framework developed in this volume addresses.

Transport

The main transport policy objective, both at member state and EU level, is to ensure there is sufficient mobility to support commercial transactions at all levels, that is, locally, nationally, intra-EU and extra-EU. Intra-EU trade, in particular, has grown in line with the evolution of the single market and the single currency. Failure to develop the infrastructure that supports this additional cross-border trade undermines the efforts to develop the single market. Bottlenecks, congestion and inadequate links between the periphery and the rest of the Community continue to dog Europe's transport networks. In addition to the general obstacle that transport networks have traditionally been designed and developed within national borders and without too much consideration of the benefits of transnational interconnection, Europe's transport networks are composed of several modes – road, rail, sea, inland waterways, air – each of which pose their own particular challenge in developing cross-border links.

Transport has traditionally been a matter almost entirely for the state. This has been for conventional military reasons – to ensure territorial security – and for financial reasons as the long-term, large-scale projects involved in the development of transport infrastructure have not been overly attractive to the private sector in terms of securing a return on its investment. Moreover, investment in transport infrastructure has often had socioeconomic objectives as a result of its role in regional development which does not sit well with private sector concerns. For many years, the state was not only heavily involved in the provision of infrastructure but it also had the provision of many transport services tied up. This was certainly the case in terms of rail and civil aviation but was also true in relation to many bus and coach services, airports and seaports. Accordingly, pre-SEM transport networks were some of the most heavily regulated services in Europe.

As ideas of liberalism and the application of these ideas spread, so the degree of state control over transport lessened. Liberalisation of road haulage and civil aviation took effect in the early phases of the construction of the single market and has wrought major changes in the structure and pattern of these sectors. However, rail liberalisation has taken much longer to achieve and, in the mid-2000s, its impact still has to take hold. In other aspects of the transport industry, notably ports and airports, the loosening of state control took the form of transferring ownership from the public to the private sector. This trend is far from universal (many ports and seaports remain in the public sector, either at national or regional/local level) but the idea of ports and seaports competing with each other has helped imbue a more commercial approach to those transport nodes remaining in the public sector. A further aspect undermining control has been finance, or the lack of it, in the public sector. This has been apparent in southern Europe where, in several states, responsibility for the construction and management of major roads has, within the context of overarching state plans and guidelines, been conferred upon the private sector.

Despite these demonstrations of liberalism and greater role for the private sector, the public sector continues to play an important, albeit more indirect, role in achieving the goal of integrated trans-European transport networks and a reduced mobility gap. For example, one of the main requirements of developing transnational transport networks is interoperability: the European Union and member states have been instrumental in driving this forward where appropriate. Interoperability is crucial in a technical sense, especially but not only in the rail sector, but the state has also in other aspects, played a key role in improving interoperability – for example, in relation to qualifications and ensuring access to professions (see Chapter 3 in relation to road haulage, for example). The public sector also continues to play a regulatory role in ensuring that the prevailing business conditions encourage the greater involvement of the private sector in the development of Europe's transport sectors: that is, in facilitating market access and making sure those obstacles to common transport markets are removed.

Overcoming the mobility gap entails all the usual network sector problems of liberalisation and of guaranteeing access to transport markets in other member states, both by regulation and by the construction of the appropriate infrastructure. However, the transport sector is made up of several transport modes. One of the challenges for transport policy makers is not only to facilitate links and connections within modes but to promote connections between modes.[1] Successful intermodality will play a major role in reducing congestion by diverting traffic to other transport networks. Some progress has been made in this direction but much more is possible, although the Commission's targets for shifts to rail, whilst desirable, may be over-optimistic.

Telecoms

The policy-strategy interface within the telecommunications sector is driven by the political desire to stimulate the information society within the EU. In policy terms, this objective is based on the deployment of ICTs throughout Europe's socioeconomic structure. Many of the EU's policy documents (as well as those of member states) have put the development of the information society as the centre of its growth strategy, a prominence that has been especially evident since the 1990s. Policy makers tied the evolution of the information society to the maturing of market forces with the creation of a single European information space based on interoperable and integrated networks across a common service environment. It is evident that the initial process of liberalisation stimulated strategic change within the sector, provoking the need for further policy responses. The liberalisation of the sector promoted change in terms of service environments as new entrants and new technologies emerged to challenge the incumbents.

With the extension of the final phase of liberalisation to infrastructure itself, the intention was to transform the integration of the service environment into an integrated series of networks. This phase of the liberalisation process sought to expose the remaining aspects of the incumbent to full competition. The end result was to develop a pluralist system of networks across the EU to support the emergence of the information society. By and large, it can be said that the liberalisation actions have improved and aided the development of the European information society. Broadband and mobile technologies (two key drivers behind the information society) have spread throughout the EU. This has been supported by heavy investment in transnational backbone infrastructure across the continent.

However there have been constraints upon and hindrances to the ability of the liberal route to develop the information society in the manner envisaged by the European Commission. A major problem is that a key part of the infrastructure remains under incumbent control. The local loop is central to gaining access to the end-user but local incumbents have restricted access with the result that investment in this aspect has been curtailed. Whilst over the long term, this obstacle is being surpassed by new technologies, it has – over the short term – represented a significant road block to the European information society. This problem has been compounded by the broader implications of the 'telecoms crash' for the development of the information society. This has been broadly felt in two areas. First, many of the new entrants left the marketplace. Second, many players (both incumbents and new entrants) scaled back their investment plans in the networks to support the information society.

To some degree, these developments have led policy makers to retreat from the ultra-liberalist stance taken towards the development of the information society since the 1990s. The bursting of the telecoms 'bubble' in the late 1990s has left the Commission (as well as national public bodies) facing a policy dilemma between persevering with the market-led route to the information society, delaying the onset of this policy objective or altering its stance to speed up the development through more strategic methods. As a result, there has been a limited shift away from the liberal stance. Policy makers have come to realise that the most effective means – over the short term at least – to realise the information society is through incumbents. These are the national operators that have the most extensive network reaching into those parts of the European economy which are likely to be underserved by market-driven forces, notably remote and deprived regions. Thus there have been some active policy measures by the state to trade-off limits to the intensity of competition with a commitment by the incumbent to invest in the infrastructure to support the information society. In other instances, states – with the agreement of the Commission – have offered subsidies to speed the roll out of broadband in areas with sparse population densities.

Overall, the interface between corporate strategy and public policy in telecommunications has been redefined several times over the past few decades. The traditional development paradigm has been replaced by a broad policy agenda that has linked network development to enhanced freedoms for corporate strategy. There is less effort by governments to control investment. The regime change has had a profound impact on all aspects of the network environments. Liberalisation has generated innovation in services that has stimulated increased demand for network capacity. This, combined with broader economic integration, has generated the demand for cross-border infrastructure. However the process of regime change has been curtailed by the sustained importance of incumbents. These have deployed effective non-market strategies to forestall the intensity of competition in recognition of their prominent role in creating an inclusive information society. Thus, despite the shift towards liberalism, such movement remains incomplete. Indeed in some instances (such as the commodity elements of network capacity), there has been a partial reversal as consolidation occurred.

Energy

No basic task can be carried out without energy, including transport and communication, the activities inherent to the other TENs sectors. Public policies towards energy at both European and national level have for many years focused on one or more of the following three objectives – liberalisation and the development of competition within individual energy sectors; security of energy supply – that is, ensuring reliable and suitable supplies of energy at all times – and environmental issues. The predominance of each objective depends on prevailing economic and political factors. For example, after the oil price hikes of 1973 and 1979, security of supply was the dominant objective of energy policy – and, as part of this, states were keen to retain their control over this sector. However, by the mid-1980s, real oil prices had fallen dramatically following the discoveries of new sources of oil, spurred on by high prices, and production of oil in areas where it had been previously unprofitable to exploit hydrocarbon resources. Thus, for the following few years at least, security of supply concerns were rarely voiced.

Since the mid-1980s, liberalisation and utilisation of the market have gradually become more important for energy in Europe. The enthusiasm for surrendering the old control models was greater in some countries than others and, as Chapter 5 shows, incumbents in some European states have managed to hold onto much of their influence as their governments demonstrate reluctance to give up some, if any, control. This has, however, had unfortunate implications for other member states where liberalisation has been respected in practice as well as in law. Utilities from these states frequently have to compete in their domestic markets with utilities from other countries but are unable to compete in the markets of all their competitors because of a lack of reciprocity in relation to market access. Moreover, utilities in the more liberal member states have been restructured prior to liberalisation which usually means they have been broken up to foster competition in the domestic market, thereby depriving them of some of the critical mass needed to prosper in a marketplace spanning several countries.

Whilst the liberal approach has been gathering pace, security of supply has moved up the energy agenda once more. This time, security issues are more complex and are likely to stay around longer. Previous supply scares have centred on the supply of one fuel – oil – as a result of uncertainties in one producing area – the Middle East. This time, supply concerns are not only about oil but also about natural gas and, in some scenarios, potentially even about coal. Supply concerns stem not only from geopolitical disruptions, although these remain a very real part of the supply security equation, but also from the decline of indigenous production of natural gas in the European Union and from greatly intensified competition for supplies in international markets as a result of the rapid demand growth in developing countries like China and India. These supply concerns affect natural gas directly but also electricity because, as a secondary energy source, it needs to secure its fuel inputs from somewhere. Increasingly, European generators have turned away from coal towards gas as their primary energy input.

So far, greater security concerns have co-existed with the liberal philosophy. Indeed, the supranational perspective tends to portray the liberalised market as going hand in hand with greater security of supply. This scenario requires a large degree of solidarity among member states and, at a practical level, assumes that greater interconnections will help energy supplies reach those parts of the Community in short supply. This may work in a sudden short-lived supply crisis but states prove highly reluctant to surrender control when their fundamental interests are at stake. As such, unless all member states can be persuaded there is an overwhelming case that their best interests are served by relinquishing control, there is a big probability that in the case of a sustained supply crisis member states will seek to reverse some of the recent changes in an attempt to regain control over their utilities and energy supplies.

* * *

In Chapter 1, a framework was offered to contextualise the shifting nature of the policy-strategy interface across Europe's network sectors. The framework suggested that a process of change was underway as the Commission and national policy makers shifted towards market-based provision. One implication of the research undertaken for this book is that the trajectory of change within these sectors and across has been uneven, even if the commitment to the process of market opening largely remains. There are several interlinked explanations for this unevenness across states and markets.

The first is that one impact of the process of change and market opening across the network sectors has led to a disaggregation across most of the respective sectors.[2] This has happened both vertically and horizontally. Vertically, the integrated industry has been broken up with specialists emerging within and between each state. A consequence of this vertical disintegration has been that the pace towards liberalisation has been uneven with the more market-driven elements progressing towards the liberal model whilst the rump of natural monopoly elements have tended to lag in the face of the intensity of competition. Thus for incumbents different aspects of their business have headed towards a more liberal business faster than other elements.

Vertical disaggregation has in some cases been accompanied by horizontal segmentation. In the more liberal element so the marketplace many firms have sought to challenge the incumbents by targeting specific groups of customers within a specific stage of the industry value chain. Naturally, as the new entrants target the most profitable segments some of the least profitable customers have not felt the benefit of the shift towards the liberal regime. Thus in these cases choice remains limited and service quality curtailed and investment muted.

A second factor that has driven the unevenness of regime change has been the nature of the policy actions taken by the respective states[3] and even by the Commission. A common factor across all the network sectors is that their social and strategic functions extend beyond pure market-based provision. As a consequence, for political and social reasons, states sometimes decide to implement the process of regime change in a protracted manner. In part, the reason for such a policy stance could be down to cultural factors such as a lack of confidence in the ability of markets to deliver networks that are socially desirable and politically expedient. This has had its most potent expression in the approach of liberalisation to essential facilities. In many cases, policy makers have resisted the speedy liberalisation of this element in order to sustain what they see as key priorities.

The third factor is the effectiveness of non-market strategies to delay or resist the onset of competition within key parts of business. Broadly, the more policy interests are aligned to those of the incumbent, the more non-market strategies are likely to be successful. Generally it is believed that the history of the incumbents as former state owned businesses has enabled them to build relationships with political authorities and national institutions that could make policy makers more responsive to incumbent lobbying. The emergence of independent regulators was supposed to reduce the threat of regulatory capture by incumbents but the potential for incumbents to resist change, with either the tacit or explicit support of official organisations, is still present and exploited.

Regime change in Europe's network sectors: the shifting policy-strategy interface

Observing commonalities of change across diverse sectors with increasingly disintegrated value chains is difficult. From the above, it is evident that the policy-corporate strategy interface across the sectors differs according to the nature of the perceived problem or the broadly defined socioeconomic outcome sought.

Nodes

Nodes play an important role in transport because it is through nodes that users of transport can gain access to each mode – e.g. access to rail transport is only possible *via* rail stations for passengers or through stations or other nodal forms for rail freight. In road transport, particularly motorways which form the basis of trans-European road links, nodes take the form of motorway junctions or access points to the links. Nodes are also important because it is through nodes that intermodalism develops. Distribution centres (the key to logistics revolution), for example, often frequently develop around ports, airports and where there is access to road, rail and inland waterway networks.

Public policy interest in nodes is not uniform across all sub-sectors in transport. Generally it is through aviation and shipping (modes in which it is the nodes rather than the links that determine the efficiency of the network and its connections) *via* the development and construction of ports and airports that nodes have exercised the attention of policy makers. Although many seaports and airports remain in the public sector, growing commercialism and private involvement in nodes like airports and seaports reflects an easing of control. However, the public sector tends to retain a major interest in nodes because of socioeconomic and regional development objectives.

In telecommunications, the major debate concerning nodes has centred on access to local exchanges as part of the process of local loop unbundling. The local loop is a pivotal part of the telecommunications networks as it provides direct access to the end-user by the supplier. By and large, incumbents have attempted to forestall access to these local exchanges by incumbents. The excuses offered for these delays have varied from logistical problems to a belief that the process is simply not needed due to technological change. The latter point seems especially pertinent as the development of mobile technology and new technology such as wi-fi are gradually diminishing the need for access to local exchanges for the purposes of delivering broadband services. Mobile broadband and its spread is undermining the need for liberalisation of this node. Thus, the local loop is only one method through which universal access to broadband will be achieved. Furthermore, it is likely that access to broadband *via* this method is likely to remain dominated by incumbents.

Overall, despite the sustained dominance of incumbents within this aspect of nodal investment, the emergence of new technologies for the delivery of broadband highlights that there are few problems garnering the necessary investment into this aspect of infrastructure. This is compounded by the fact that there is no shortage of potential investors wishing to put their own equipment into local exchanges. Thus, the natural monopoly element has been eroded and it is only the reluctance of the incumbent to see the remaining elements of control eroded that has limited its impact. However the telecoms crash has had some tangible impact on nodes as the financial aftermath of the crisis led to many operators scaling back their plans for the roll out of advanced mobile services – a key platform for the development of the information society.

In the energy sector, the emphasis is not on constructing or building new nodes but on the regulatory issues that ensure nodal owners and operators can gain access to the key links to ensure they can transport product to their customers. For electricity, nodal requirements tend to focus on nodal (i.e. power station) operators gaining access to transmission grids so that electricity can be transported to the customer. Gas producers also need to secure access to networks to facilitate the transportation of gas. In the case of LNG, producers also need access to gas liquefaction and gasification plants to transport fuel *via* carriers and to get access to pipelines from the port to the end-user. LNG has historically been a relatively small proportion of the gas movements into Europe because of relative costs though there are signs that it is becoming more important as it loses some of its cost disadvantage. If LNG does become more important, the rigidity of gas networks will be reduced as the sourcing of product will not be limited to the configuration of pipeline network.

Services

Liberalisation has been at work in all transport sectors, albeit on different time scales. Service provision occurred first in road haulage and airlines. Rail liberalisation is still underway in the mid-2000s and its effects have still to become fully evident. Over and above sectoral specific liberalisation, the market has become a more powerful driver in services. This is particularly the case in the convergence among service providers with origins in different sectors – many operators originally in shipping or road transport, for example, have reinvented themselves as full-service logistics providers with capability to deliver goods efficiently through their seamless use of different networks. The origins of this switch pre-dated the birth of TENs and the onset of liberalisation but these two events have both accelerated and pushed this trend forward.

By the turn of the millennium, telecommunications within the EU had been completely liberalised. Whilst the approach was phased with some states reaching full liberalisation before others, the laggards generally felt there was little incentive to forestall the process and limit the opportunity for investment. The new regulatory framework established within the Electronic Communications Directives has sought to account for the commercial and technological convergence between telecommunications and the IT and content sectors and has updated regulation to allow for the increased flexibility of business models. Over time, it is anticipated that the need for sector specific regulation will become redundant as competition matures and the normal rules of competition can be applied.

However despite the commitment made by states to the liberalisation of services, implementation has been uneven across the EU. This irregularity could be due to reluctance by member states to undermine incumbents or to effective non-market strategies by incumbents. This uneven pattern has continued into the Electronic Communications Directive where implementation of certain key aspects has yet to occur. Thus, despite a great deal of progress in the development of liberalisation across the EU, further effort is still needed to integrate the marketplace fully.

Services have been liberalised in both TENs energy sectors but the impact of liberalisation has been variable throughout the EU because of differences in how member states have interpreted the relevant directives. Some markets are completely open and competition is intense and shaping the industry whereas in other member states incumbents have managed to preserve their dominant position and limit the effects of liberalisation.

Links

Links are not important for air or short sea shipping (what matters for them are nodes) but are crucial for road, rail and inland waterways. In these latter sectors, the challenge is to develop new links – a process which can take a long time given the often complex and drawn out authorisation and planning processes – and to improve the quality of existing connections. This includes the upgrading of rail lines and roads as well as developing new features on these links such as bridges (e.g. the Oresund and Fehmarn links) and tunnels to overcome bottlenecks or link two previously unlinked locations.

Substantial investment in pan-European telecommunications links occurred throughout the 1990s. By the beginning of the new millennium, competition in the pan-European transmission market had intensified following liberalisation and the expectation of rising traffic had attracted substantial investment funds. However there is an argument that the amount of funding attracted into the sector led to unstable market structures which pre-dated the market crash in the closing years of the twentieth century. Despite this market instability, Europe still had the necessary capacity within its backbone to support the emergence of the information society.

However problems remain with links elsewhere within the network. First (as mentioned above), roadblocks still remain within the local loop infrastructure. The links between the user and local exchanges need to be upgraded to cope with the emergence of the information society but (as mentioned) local incumbents have often blocked this investment. Second (again as mentioned above), the aftermath of the telecommunications crisis has seen investment in infrastructure scaled back. This has meant that not only have operators left a lot of network capacity 'unlit' but they have also scaled back investment plans in other core information society technologies (such as advanced mobile).

In the energy sector, there have been incremental improvements in electricity links. The biggest problem has been securing the authorisation needed for cross-border transmission lines: these problems have occurred because of the plethora of different planning procedures within Europe and because of often fierce local opposition to the alleged environmental impact of new high voltage lines. The big challenge regarding gas links is to secure the infrastructure to bring gas into Europe. This raises major geopolitical issues about routeing, especially regarding transit through Russia. The Ukraine gas scare in early 2006 during which European as well as Ukrainian gas supplies looked insecure has cast doubts over over-reliance on Russia. This suggests that despite the evolving impact of the market in providing links, its ability to do so can often be undermined by a broader set of strategic objectives.

* * *

Looking across the different network value chains, it is evident that liberalisation has generally initially focused upon services. Such actions were based on a premise of promoting regime change to ensure that existing infrastructure is used more efficiently and effectively to aid the competitiveness of European business. Thus much of the initial efforts are focused on increasing the intensity of competition on existing network infrastructure. Implicit in these moves was the belief that the increases in traffic that emerged from the liberalisation of services would feed through into increased infrastructure investment either by the state, the incumbent or *via* liberalisation. It is evident that the ability of the private sector to channel funds into infrastructure varies across both nodes and links as well as between sectors. In terms of nodes, there is an evident record across all sectors of private sector investment – subject to incumbent stalling tactics – being channelled into improved nodal capacity. The issue of links is less straightforward. In telecoms and energy, private funding has been realised with the picture in transport more mixed.

Strategic issues

As market-driven processes have matured within Europe, so there have been a number of changes within the network sectors. These trends have driven new business models and strategies as well as developing solutions to the major policy issues involved in regime change. It is noticeable that these trends (discussed below) have seen a change in the treatment of infrastructure. The removal (*via* regulation) of infrastructure as a barrier to entry has led to increased competition within service provision. This competition has produced a number of innovations in the service environment.

Convergence – multi-utilities

A by-product of the liberalisation process has been the emergence of multi-utilities. These have emerged from the commercial convergence of the value chains of the respective network sectors. They can be created by the sharing of network facilities or by developing a business model built on an asset light approach. The rise of the multi-utility was a response to the need by commodity producing firms to retain customers by bundling services together. This shows that the value within the network has shifted from network ownership towards service provision. Thus there is direct effort to 'own' the customer by supplying them with as many core utility services as possible.

The multi-utility is created by the bundling of utility services from a single supplier. The aim is to use bundling as a differentiator in increasingly competitive markets. Such strategies not only enable the supplier to gain from economies of scale but also offer the potential to lock the customer into the firm's products. This offers benefits to the customer in terms of simplicity of billing, etc. Thus, for many utilities, the attraction of the multi-utility is evident. They already have access to customers and their aim is to use the liberalisation process to sell more products to these customers *via* their own or somebody else's infrastructure. Thus the core competence for these organisations has shifted away from network ownership and control towards customer management. Generally many of these businesses tend to be asset light. The largest multi-utilities tend to be within Europe where firms such as E.ON, RWE, Centrica and Suez have developed a portfolio of assets.

This shift in the source of value has led to some utilities moving away from offering utility services towards other customer centric businesses such as credit cards. The aim was to replicate the customer centred skills learnt in selling utility products to other services.

However, such models have come under attack as many felt that such actions were too far removed from the core competence of the business as selling different utility services requires a differing set of skills which many utilities simply do not possess. Thus, it was felt that diversifying into these new areas was simply too risky. A utility had to move from being an incumbent to win customers from incumbents in other sectors. In the face of rising debt and the need to demonstrate shareholder value, many firms began to move away form the multi-utility model to refocus on their core business. Centrica for example reduced its diversity in the face of rising costs in its core energy business. RWE, although containing its dual role in power and gas, is reducing its presence in the water sector. In addition, Enel has also shifted away, finding it difficult to leverage its resources across multi-utilities.

Technological convergence

Alongside the process of commercial convergence has been the technological convergence between the respective sectors. The overlap between these sectors has generated the desire for more effective management of flows or, *via* a desire to sweat existing assets further, the ability for the firm to enter a new marketplace.

Telecommunications and transport is one area where there has been technological convergence of infrastructure. The interface between transport and telecommunications is dual-faceted. Transport operators have used existing infrastructure to offer telecommunications services. In a pan-European context, the clearest example of this was the emergence of Hermes in the early 1990s. This was a pan-European network based on deploying telecommunications infrastructure in the ducts alongside the rail network. Similar piggybacking in telecommunications has also taken place within the energy sector where telecommunications infrastructure has been deployed alongside the energy network.

The technological convergence between energy and telecommunications has had its most evident expression in the development of power line technology. This is a technology that uses the power distribution grid for the simultaneous transmission of data. This is at a higher speed than those conventionally used by power companies to transmit data over their networks. The adoption of power line technology looks attractive in aiding the cheap and easy spread of broadband across the EU. Indeed the technology is being touted by the European Commission as a means of offering last mile access at high speeds. However, its ability to take on and supplant existing technologies is limited (at least over the short term) due to its commercial immaturity but the fact that powerlines go everywhere and existing broadband connections do not, could make it a suitable method for extending the reach of advanced telecommunications services.

The roll out of this technology – despite its obvious attractions – has been curtailed by the fear that it will interfere with other spectrum users. However many argue that such concerns were based on early versions of the technology and that they have since been allayed as the technology has moved on. However the regulatory environment regarding the technology and its emissions needs to be clarified. This has proved difficult as there are differing opinions on the acceptable levels of emissions.

At the policy level, technological convergence has been most evident between the telecommunications and transport sectors. As highlighted above, there is evident overlap between telecoms and the Commission's desire to meet the needs reflected within the mobility gap. The EU has put substantial funding into applying ICTs to solving traffic management issues in terms of solving congestion and improving safety on road networks. Such priorities have also been evident in terms of air travel where information has become increasingly salient in integrating European skies. This convergence has also been evident in terms of the Galileo project. This project reflects a convergence between telecommunications (*via* satellite) and transport to aid navigation and positioning. This is expected to aid the efficient management traffic across multiple modes.

The rise of the virtual operator

One consequence of the vertical disaggregation of the network sectors and the horizontal segmentation of each stage of the value chain has been the emergence of virtual operators. The advent of market forces within the network sectors has led to the commoditisation of network capacity with many operators placing increased emphasis upon customer service and the superior knowledge of suppliers. The result has been the entry of firms into the marketplace that are infrastructure-based but asset light. These are businesses that provide network services but do so independently of the ownership of network infrastructure.

Such trends are evident within telecommunications where substantial investment in network infrastructure has commoditised capacity. This has led to the emergence of niche operators which build strategic networks to establish a list of suppliers to build virtual networks for customers. Such business models have also emerged within the transport sector: in logistics, firms use trusted third parties networks to deliver customer freight solutions and there has, for example, in some countries been a separation between the ownership of the rail network and the provider of services on that network. In energy, virtuality has become a feature of mature liberal markets in which gas and/or electricity suppliers piggyback on third party infrastructure to access customers.

These developments are important as they divorce service provision from network ownership. This substantially reduces the barriers to entry and removes key impediments to competition within the respective sectors. Whilst firms would always have had to rely on third party infrastructure for some part of the service delivery, these developments demonstrate that there are new ways of realising the respective strategic priorities for the network sectors. Value is increasingly divorced from network ownership and is found more in the knowledge of customer/user needs. The implication is that the key policy agendas can be enhanced by an environment that stresses competition in services.

Policy issues

In the 1980s, the SEM supported by the TENs initiative and later by the single currency was developed by the EU to bolster the flagging international competitiveness of European business. The Lisbon strategy was launched in 2000 for similar reasons, and relaunched in 2005 with the ambitious aim of creating the world's most competitive economy by 2010. The overarching themes of the strategy are growth and jobs aligned with sustainable development.

In order to achieve the Lisbon goals, actions are proposed in a number of key areas, including free and fair trade, an effective single market, improved regulation, a greater emphasis on innovation and research and development, the creation of a more adaptable, better educated and skilled workforce and an improved European infrastructure. That infrastructure should be central to the strategy should be no surprise given its intimate relationship with most of the other issues: that is, its role in facilitating the implementation of the SEM and in promoting trade; its link with innovation and research and development which contributes to bridging the mobility gap and the contribution of telecommunications infrastructure to the information society.

In most areas, the rhetoric surrounding Lisbon remains liberal (see Table 6.1). Indeed, at one level, the Lisbon process persuaded many sceptical governments of the need for reform through market liberalisation and the small print of the strategy is comprised of individual initiatives that contribute to the overall objectives of making the EU a good place in which to do business. This is consistent with the more passive approach to industrial policy outlined in Chapter 2. In practice, both before and after Lisbon, reform has not always been as forthcoming as promised by the rhetoric, both in terms of agreeing policy in line with liberal principles and particularly in terms of its implementation by member states.

In transport, mainstream liberalisation has been achieved in all sectors, albeit at different speeds with rail bringing up the rear. Other measures are envisaged that will both support the deregulated transport system and the objectives of achieving an integrated transport infrastructure. These include reform of air traffic control systems within Europe and support for Galileo, Europe's satellite navigation system, which will help foster efficient transport management systems. Work on developing a proper system for charging for the use of transport infrastructure also continues. At member state level, the emphasis is on completing the 30 transport TEN priority projects and establishing systems for electronic fee collection.

Overall, the transport policy stance is thus less one of ownership and control, especially in service delivery, than it has been in the past. However, ownership and control remains important in some modes in some locations. Nevertheless, the public policy emphasis is on measures to help transport systems operate more efficiently and effectively. On the whole, transport policy has become much more passive. The aspiration to achieve a modal shift away from road to rail and inland waterways would appear to contradict this but the strategy adopted to achieve this is one of encouragement through changes in the transport environment rather than one of compulsion.

Of all the sectors, telecommunications has tended to generate the greatest uniformity of purpose over liberalism and where to some extent corporate and political strategies have demonstrated the greatest degree of unity. However, this should not be taken that the liberal agenda is entrenched despite over two decades of liberalism. It is apparent that in the aftermath of the 'telecoms crisis' in the late 1990s, there has been a limited retreat from a purely liberal line. Whilst liberalisation remains the dominant theme, there has been a growing desire to ensure that the information economy is achieved quickly. This has often meant using incumbents as the primary deliverer of information

Table 6.1 Summary of policy stances in TENs sectors

Transport	Remains generally liberal in outlook. Most policy measures designed to facilitate more efficient and sustainable management of Europe's transport systems.
Telecommunications	Liberalisation still dominant but limited retreat in some areas for pragmatic reasons – that is, a recognition that incumbents may be in a better position to deliver the infrastructure for the information society.
Energy	Liberalisation the dominant theme – but policy not always fully implemented. Emphasis remains on completing liberalisation – through the courts if necessary.

society services which can hinder the realisation of the liberal paradigm in telecommunications.

Liberalisation remains a major theme, at least at European level for the gas and electricity sectors. The main problem has been that, although the legal framework for liberalised energy sectors is in place, several member states are maintaining a control stance in these key strategic sectors. This manifests itself either in inadequate implementation of the key liberalisation directives or in government actions to retain the position of national champions (see Chapter 5).

The Commission has so far shown a determination to take on the member states to bring about the regime change which it believes will better serve its overall energy policy objectives of energy security, competition and sustainability. As such, in June 2005, it launched an inquiry into competition in gas and electricity markets, has started proceedings against member states for failure to implement directives properly and is taking action against individual member states that attempt to block mergers and acquisitions of domestically owned utilities by utilities from other member states. In 2005 and 2006, the Commission has made increasingly determined statements about the need for open markets. The cost of defeat in this strategy will be high but success will facilitate the creation of energy TENs.

The global dimension

As the sectoral chapters have highlighted, there is a high and rising interdependence between the development of networks at the regional level and the emerging needs of the global economy. In this context, it is increasingly important to see the development of TENs as part of a wider series of sub-networks within an emerging series of global networks. In short, corporate and political strategies need to take on board the fact that their European actions are part of a wider global picture. Clearly issues surrounding energy supply, the information society and logistics are global and not merely confined to the European continent.

Corporate

At the corporate level, the globalisation of telecommunications has been driven by the emergence of a global customer base. As globalisation has evolved, so MNCs require communication networks that meet their needs across all locations in which they operate. This need, combined with a trend towards outsourcing by MNCs of these non-core functions, has opened up the market for globally managed services. The result has been attempts by many European firms to develop advanced networks in the major economies throughout the global economy where MNCs are focused. This can be achieved through a mix of developing their own assets or of using third parties. Network externalities suggest that the companies with the largest networks who are able to offer uniformity of service over this network should be able to develop the more sustainable competitive advantage. In other areas, mobile firms have also established international presence on a country by country basis. These corporate actions have an impact on the development of the European information society through the investment by foreign operators in European states and by the investment and experience gained by European operators in non-EU markets.

Global strategies in the energy sectors are more common in gas than in electricity sectors. Gas producers have long tended to have a more international approach because of the need to produce and acquire gas – activities that can only occur in gas-rich locations which in many cases are distant from the consumption areas. The challenge of transporting gas across long distances has drawn gas companies into controversies about pipeline routeing and into considerations about ensuring the risk of developing gas resources can be covered by securing markets, which in turn involves the development of networks further downstream. Given the decline of indigenous European gas resources, the engagement of European suppliers on a global stage will have to increase and/or they will be subject to enhanced competition from foreign suppliers like Gazprom. In addition, market liberalisation in both the gas and power utilities sectors has spread beyond Europe and has taken root, with varying degrees of intensity, in many of the world's regions. Assuming there are no reversals to these policies, which is far from certain, entry into markets outside Europe by European-based utilities will become more commonplace. Network interconnections, especially between smaller networks, will be sought to facilitate the reaping of synergies and economies of scale. European utilities will gain from their experience of cross-border connection and of entering new markets, either as virtual or actual operators, within Europe.

In the transport sector, global strategies are well-developed. This is especially evident in the service dimension where there are a plethora of global freight operators. Some of these have extensive global links such as those in the freight and parcel sector (such as DHL and FedEx) whilst others have a regional focus. This is also true of global shipping and air transport, where access to key links is improving and enabling the spread of global carriers. These global carriers are the most evident expression of where corporate strategy is creating global mobility for firms and consumers (at a price). This trend is, in part, a response to the demands of customers who, either through practices such as outsourcing, just-in-time management or the development of transnational value chains, are requiring the evolution of transnational transport and distribution services to reflect their own transnational dealings.

In the transport and telecommunication sectors, the global element of the networking environment has had its most potent expression in the emergence of global carriers. As suggested they reflect the fact that mobility, whether physical or virtual, is a strong need for some customer groupings. These concerns do not necessarily reflect policy concerns but they can exercise an influence over them. Clearly the ability of these global carriers to offer this global product is based on the spread of the liberal agenda across the network sectors.

Policy

Just as the integration brought about by the SEM contributed to the rationale for developing trans-European networks – that is, in support of the increased economic interdependence and transactions brought about the reduction of trade barriers within Europe – so a similar, albeit more spontaneous, process is at work in relation to international networks. Globalisation, or increasing international integration, like European integration includes lower barriers to trade and investment. In order to maximise the benefits from this, as with regional integration, physical networks are needed to support this additional economic activity.

The options for Europe in breaking out of its territory in terms of infrastructure vary according to sector and geography. In the energy sector, the emphasis is on developing new links. In the electricity sector, there is much discussion of developing a Mediterranean Ring (incorporating the countries of North Africa and the eastern Mediterranean) and early preparations for this are underway. Possible extension of networks into south-eastern Europe, where the next wave of enlargement is likely, is also under consideration. These initiatives potentially represent an extension of the SEM and are intended to improve energy security across a wider area. Power links with Russia, which has an excess of power, are also on the agenda in the medium to long term, are regarded as a way of improving security of supply. In order to maximise the benefits from any expansion of the reach of power grids beyond Europe, greater interconnection needs to be accompanied by liberalisation in the outreach areas. At present, such initiatives are been driven by state or regional actors rather than by the private sector. However, if the extension of links beyond the EU's borders does occur in any major way, then existing organisations that exist to facilitate the practical aspects of interconnections, such at ETSO (the European Transmissions Systems Operators organisation), the European Regulators Group or the Florence Forum are likely to expand their remit and membership.

In the gas sector, although improvement of the domestic network is important, the decline of indigenous production coupled with growing demand for gas means the greatest imperative in relation to gas is to bring more gas into Europe from major suppliers. This entails the construction of long, high pressure pipelines from Russia, Central Asia, the Middle East and North Africa plus construction of gas liquefaction and gasification plants to bolster LNG. The EU has been developing its own energy diplomacy to help this process along. In particular, it has opened an Energy Dialogue with Russia to help it secure supplies of Russian gas. This link has not been without its problems which stem from a number of causes. However, the important thing to note is that the securing of gas supplies has become such an important issue for Europe that the involvement of non-commercial organisations like the EU and member states' government are inevitable.

In both energy sectors, the policy involvement comes not through multilateral organisations but is largely bilateral. This is not entirely the case in the transport sector. For many years preceding the formation of the EU in the 1950s and the even later beginnings of a Common Transport Policy in the 1980s, there has been international involvement in the regulation of airline and shipping, in particular. The International Maritime Organisation, for example, is a UN body. It deals primarily with safety and technical issues. Regulatory issues at an international level affect certain trades only and tend to be dealt with by organisations like the OECD and UNCTAD. Aspects of the activities of international airlines have been regulated by the Warsaw and Chicago Conventions of 1929 and 1944 respectively. Regularly updated by the International Civil Aviation Organisation, the Chicago Convention is primarily concerned with safety, technical issues and the professional qualifications and standards of airline personnel. However, organisations like the WTO, and before it the GATT, which are concerned primarily with access to markets have not played a role in the liberalisation of transport services. As tariff barriers have reduced and services have been thrust more into the WTO spotlight, the possibility of liberalisation of transport services on a global basis *via* the WTO has been raised but little has been done to achieve this – and the prospects for it look slim in the summer of 2006 given the suspension of the Doha Development Agenda and the inevitable soul searching that will take place about the role of the WTO itself.

In the absence of a widely agreed international forum designed to provide the integrated transport system which the increasingly integrated world economy needs, the emphasis on opening up transport systems internationally is falling upon bilateral initiatives. After much wrangling with member states and with the assistance of the European courts, the European Commission began negotiations on an aviation agreement with the US in 2003. The EU has managed to reach bilateral agreements on airlines services with several states but the agreement with the US, the most important agreement for the EU to reach, is proving more problematic. In short, in order for transport networks to follow developing patterns of trade and economic interchange, an appropriate framework of multilateral institutions does not exist, although institutions do exist which could evolve into the role. For the time being, however, the policy-strategy interface for the extension of transport networks beyond Europe, consists of a combination of bilateral negotiations in airlines and other sectors and the efforts of international logistics operators, infrastructure owners and transport users to develop the appropriate services and networks despite the inadequacies of the policy-making institutions and to use their influence with national and European organisations to respond to these economic pressures.

Policy frameworks within global telecommunications are fairly well-established. The Commission has worked within the ITU to aid the establishment of global standards. The development of the global information society has been aided by the open nature of Internet standards which have enabled interoperability. Interconnection has also been fostered by the WTO agreement within the GATS. This agreement which came into force in 1998 set an agenda for the opening of markets across 69 states covering 90 per cent of traffic. As the EU had already committed itself to a liberalisation process which superseded the commitments within this agreement, this had little direct impact upon EU states other than their relations with third parties.

The interconnection and interoperability of networks on a global basis reflects the fact that the European Information Society is part of a wider Global Information Society. Thus EU policy makers are looking to develop networks that exploit global issues in a policy that seeks to develop the information society. The World Summit on the Information Society launched in 2001 was a formal UN initiative to build a global approach on the information society. These reflect a broad pro-market-driven information society at the global level.

Conclusions

Whether it occurs on regional or a global basis, internationalisation is ultimately about mobility. The importance of infrastructure in terms of international markets reflects the need for the mobility allowed by shifts in policy to be facilitated by the supporting infrastructure. This has meant the creation of a series of supporting infrastructures. At the European level, progress has been mixed but there are grounds for optimism as environmental contingencies promote further shifts in the policy-corporate strategy interface. However as this chapter has concluded, this is only part of the story as these trans-European networks have to be seen as part of a sub-network of an emergent global network of networks. As European business starts to look beyond its home region so the salience of the absence of these networks will start to emerge as a policy issue.

Notes

Chapter 3 Transport TENs

1 European car ownership levels lag far behind US levels of 769 cars per thousand inhabitants. Convergence towards US levels would only intensify Europe's congestion problems.

2 In the 1960s, pipelines were regarded as a mode of transport and were included in discussions about transport infrastructure. The pipelines referred to include the possibility of 'the development of pipelines for the development of mineral oils' (CEC, 1960: 188). Over time, gas pipelines grew to be regarded as an energy infrastructure issue but oil pipelines have not been included in the energy TENs initiative, although in the early to mid-2000s the incorporation of oil pipelines into TENs has been aired.

3 Progress was so slow that the European Parliament took the Council of Ministers to the European Court of Justice in 1982 for failure to fulfil its Treaty obligations to formulate a common transport policy. The Court decided in favour of Parliament in May 1985.

4 Despite the Christophersen Group's insistence that its work does not duplicate that of other TENs actors, it has encountered fierce criticism from the European Parliament which claims that the intergovernmental process circumvents the procedures laid out in the Maastricht Treaty.

5 The same general criteria used in selecting the 14 priority projects were also used for these 21 projects 'of further importance' but the projects themselves were not as far advanced as the 14 priority projects.

6 The three were the Oresund fixed link between Copenhagen and Malmö; Malpensa Airport in Milan and the Cork–Dublin–Belfast–Stranraer rail line. In the case of the latter, some capacity enhancement had been proposed and was included in List 1 as a new project.

7 A motorway requires three times the quantity of land than the equivalent length of rail track.

8 These first TSIs were concerned with six subsystems – maintenance, control and command, infrastructure, energy, rolling stock and operations.

9 NAIADES = the Navigational and Inland Waterways Action and Development in Europe and is a reference to fresh water nymphs in Greek mythology.

10 In the 1980s, for example, a large part of the UK port sector was privatised and formed into Associated British Ports; the 1991 Ports Bill made possible the privatisation of the larger trust ports (i.e. locally owned ports). Independent private ports and wharves continue to exist. Ownership patterns are equally varied throughout the rest of Europe.

11 These include the old-fashioned image of shipping; the complexity of the administrative procedures required to support the passage of freight through a port; the absence of an efficient door-to-door element for shipping services; the need for improved and cheaper port services such as

pilotage, towing, cargo handling and services provided to passengers; and issues that are specific to individual countries. In 2001, the Commission proposed a directive on access to port services with a view to increasing their efficiency and lowering their cost but it failed to progress because of opposition by the European Parliament.

12 According to the Commission, it is intended that a number of specific projects will be identified by early 2007. In order for this to happen, some of the difficult decisions alluded to will have had to have been taken.

13 Axes and projects involving other modes do have the potential to increase the accessibility of airports.

14 Galileo is a European initiative to create global satellite navigation systems and is one of the 30 TENs priority projects. The EU has also supported in different ways the following traffic management systems to help traffic flows between member states: the European Rail Traffic Management System (ERTMS); River Information Services (RIS); Automatic Identification System (AIS) and Long Range Identification and Tracking (LRIT).

15 For example, London's Heathrow Airport was built on the site of a former privately-owned aerodrome which was requisitioned by the Air Ministry in 1944 to become a major base for the RAF. This trend continues to this day. One of Europe's newest international airports Robin Hood Airport Doncaster Sheffield (opened in April 2005) is located on the site of RAF Finningley which was established as a military airfield as far back as 1915, a status which it retained until 1996. The length of the runways used by the RAF are enabling the new civil airport to rapidly develop its international services.

16 For many years, the existence of a national flag carrier was regarded as a vital symbol of national prestige. However, many of these airlines were unprofitable and required sizeable subsidies. This was incompatible with a single aviation market. Although state aids were a major problem when the liberalisation process was underway, they have declined significantly since then. The bankruptcy of the Belgian national airline and the effective takeover of the Dutch flag carrier, KLM, by Air France symbolises the decline in this particular political role of airlines.

17 Anguera (2006) argues that the UK economy would have been better off if the tunnel had never been built because the total resource costs outweigh the benefits. In short, the investment has poor viability financially and in cost-benefit terms.

18 At the time of writing, it is too early to say whether this issue will flare up into a major political confrontation, but it does indicate, even within a country that has adopted some of the most liberal positions in relation to transport, that control instincts and power of national symbols remain.

19 A classic example of a political reason to embark upon a major infrastructure project was the decision in the 1960s to proceed with construction of the Humber Bridge. The deciding factor in the decision to proceed with the project was the imminence of a key by-election in Hull, at the northern end of the route of the bridge. If the government were to be defeated in the by-election, its majority would have disappeared.

20 Estimates of the market share of 3PLs vary. In 2000, Carbone and Stone (2005) reported Datamonitor estimates that 3PLs contributed to 26 per cent

of total logistics provision in the EU. Developments since then suggest that the market share of 3PLs has increased significantly.

21 The company rebranded itself to DSV from 30 May 2006.

Chapter 4 Trans-European Telecommunication Networks

1 A gigabit per second (Gbps or Gbit/s) is a unit of data transfer rate equal to 1,000 megabits per second or 1,000,000 kilobits per second or 1,000,000,000 bits per second.

2 A terabit per second (Tbps or Tbit/s) is a unit of data transfer rate equal to 1,000 gigabits per second, 1,000,000 megabits per second, 1,000,000,000 kilobits per second, or 1,000,000,000,000 bits per second.

3 Dark fibre is optical fibre infrastructure that is currently in place but is not being used. Optical fibre conveys information in the form of light pulses so the 'dark' means no light pulses are being sent. Dark fibre can refer to infrastructure that is in place but not yet ready to use.

4 Dense Wavelength Division Multiplexing is an optical technology used to increase bandwidth over existing fiber optic backbones.

5 Asymmetric digital subscriber line is a new technology that allows more data to be sent over existing copper telephone lines (POTS). ADSL supports data rates of from 1.5 to 9 Mbps when receiving data (known as the downstream rate) and from 16 to 640 Kbps when sending data (known as the upstream rate).

6 The European Competitive Telecommunications Association looks after the regulatory and commercial interests of new entrant telecoms operators, ISPs and suppliers of products and services to the communications industry.

Chapter 5 Energy TENs

1 For example, energy security issues were a major concern for some years following the 1973 and 1979 oil price hikes, both of which were precipitated by political crises. Latterly the Iraq crisis, tension over Iran and other political issues plus burgeoning demand in China and India in particular have thrust supply security issues back to the top of the agenda.

2 This may help in the case of short-term shortages but not if there was a general supply crisis. Moreover, one of the main outcomes of liberalisation is supposed to be lower prices; lower prices encourage demand growth and higher demand growth reduces supply security.

3 Oil is not formally part of the TENs programme but has been discussed within the context of TENs during the 2000s.

4 Secondary energy, like electricity, is obtained from the conversion of energy contained in raw fuels like coal, natural gas and oil – which are the primary energy sources.

5 The North European Pipeline across the Baltic Sea and ultimately reaching the UK is in its early planning stages.

6 Relevant projects included LNG terminals, energy production (electricity generation and oil refining), natural gas and electricity supply lines plus petroleum and natural gas storage installations.

7 The guidelines and lists of projects have been revised in 1997, 1999, July 2003 and, at the time of writing draft revisions of the guidelines to take into account the 2004 enlargement await adoption.

8 The ten Christophersen priority projects included electricity interconnections between Greece and Italy; France and Italy; France and Spain; Spain and Portugal; Denmark – connection of the east and west networks; development of a gas network in Portugal and a gas network in Greece; gas interconnection between Portugal and Spain and the construction of two gas delivery pipelines from neighbouring countries – one from Algeria, through Morocco and up through the Iberian peninsula and the other from Russia through Belarus and Poland to Germany.

9 The original nine were Austria, Belgium, France, Germany, Italy, Luxembourg, the Netherlands, Switzerland and Western Denmark.

10 The objective of this new priority axis is to enable the spare capacity in Central European states to be made available to other countries. In the process, this priority project is intended not only to improve links between Central Europe and Austria, Germany and Italy but to enable southern consumers to benefit from new wind generation in northern Germany and will improve the efficiency of the internal German network.

11 Gas supplies to Portugal were made possible because of the interaction of three of the original TENs priority projects in the Christophersen Report – the creation of a natural gas network in Portugal, the Spain-Portugal gas interconnection and the gas delivery pipeline from Algeria, through Morocco and up into the EU trough Spain.

12 Funding is available for feasibility studies and for key TENs projects from the European Investment Bank.

13 CEC (2004b) *Medium Term Vision for the Internal Electricity Market,* DG Energy and Transport Working Paper, Brussels, 1 March 2004.

14 Although not a member of the EU, Norway is a member of the European Economic Area (EEA), effectively making Norway part of the single market. Norway is a key component of the Nordic energy market by virtue of the dominance of its hydro generation which complements the thermal generation elsewhere in the region: that is, when water levels are low, Norway imports electricity from the rest of the Nordic countries and exports power when water levels are high. Norway's status as a gas producer also confers major importance on Norway's involvement in gas TENs.

15 The author's experience working in the UK ESI in the early 1980s revealed strong constraints on its activities. For example, if the industry's economic forecasts (on which plans for new plant, among other things, were based) were out of line with Treasury forecasts (that is, more pessimistic), then the industry's forecasts were rejected. In addition, analysts within the ESI were told to produce forecasts to justify the building of Sizewell 'B' nuclear power station, the subject of a drawn out public inquiry in the 1980s, rather than to produce independent forecasts that would determine whether the project should go ahead or not. In other words, the industry was essentially used to support the government's policy of a nuclear future rather than to develop its own policy.

16 Until well into the 1980s in the UK, for example, the UK ESI's fuel decision was constrained by the requirement to buy given quantities of coal

domestically to protect the national coal industry rather than source the cheapest supplies.

17 Although overall market concentration in generation is relatively low in Germany, the large generators retain dominance within clearly delineated regions of Germany – a legacy of the pre-liberalisation era.

18 In July 2004, E.ON signed a memorandum of understanding with Gazprom regarding joint development of the Yushno Russkoje, a gasfield in Western Siberia. Production is estimated to start in 2008 and the reserves are forecast to last 25 years. Although precise details are still to be determined, E.ON is also a participant in the Vyborg-Greifswald pipeline due to start operations between 2010 an 2012.

19 One consequence of this is that UK utilities are not amongst those driving for regional or even pan-European status. Rather their relative size has meant they have become targets for utilities from other European countries. On the one hand, it can be argued that the nationality of pan-European utilities does not matter. On the other hand, the break up of the UK ESI meant that they were never in a position to compete on the same terms in the European marketplace with utilities from larger European countries where liberalisation occurred later and differently. In short, the UK utilities were compelled to face opening of Europe's energy markets under similar circumstances to utilities in smaller member states.

Chapter 6 Themes in the Development of the European Network Economy

1 This challenge is unique to the transport sector. Although there are clearly benefits from gas-electricity links (such as bringing gas supplies into power stations), these are limited and do not contribute to the overall efficiency of the network.

2 However, there has been some vertical reintegration in the energy utilities (see Chapter 5) which has undermined some liberalisation aspirations.

3 The use of directives also opens up scope for differential regime change – a factor which is discussed in Chapter 5.

Index